115 Advances in Polymer Science

Photoconducting Polymers/ Metal-Containing Polymers

With contributions by
M. Biswas, A. Mukherjee, V. Mylnikov

With 88 Figures and 9 Tables

Springer-Verlag
Berlin Heidelberg GmbH

ISBN 978-3-662-14903-4   ISBN 978-3-540-48192-8 (eBook)
DOI 10.1007/978-3-540-48192-8

© Springer-Verlag Berlin Heidelberg 1994
Originally published by Springer-Verlag Berlin Heidelberg New York in 1994
Softcover reprint of the hardcover 1st edition 1994

Library of Congress Catalog Card Number 61-642

Typesetting: Macmillan India Ltd., Bangalore-25

SPIN: 10114940   02/3020 5 4 3 2 1 0 Printed on acid-free paper

Editors

Table of Contents

Photoconducting Polymers

Vladimir S. Mylnikov
Institute of Cinema and Television, Pravda Street 13, St. Petersburg 191126,
Russia

Photoelectric phenomena in all the available classes of organic polymers are reviewed. Photogeneration, transport and recombination processes of the charge carriers are analysed from the point
of view of semiconductor physics. Special attention is given to the relationships between chemical
structure and photoelectrical properties of the macromolecular compounds. Chemical and spectral
sensitization of the photoconductivity by dyes and donor-acceptor molecules is discussed. The status
and prospects of the application of the polymeric photoconductors as new prospective electronic
materials are analyzed in optoelectronics, holography, nonlinear elements, electrophotographic and
photothermoplastic recording.

List of Symbols and Abbreviations

$\Delta\sigma$	photoconductivity
e	electron charge
Δn	concentration of the excess electrons
Δp	concentration of the excess holes
μ_n	electron mobility
μ_p	hole mobility
ε	mean charge carrier life time
i_{ph}	photocurrent
F	generation velocity of the charge carriers
G	gain coefficient
t_{tr}	transit time
v_d	drift velocity of the charge
E	electric field strength
V	voltage
L	distance between electrodes
C	capacity
τ_r	dielectric relaxation time
ΔB	width of the photoconductor band transmission
τ_o	response time of the photoconductor with traps
n	concentration of the free carriers
n_t	concentration of the trapped carriers
v_o	surface potential
ϕ	generation efficiency
I	light intensity
CT	charge transfer state
φ_0	initial quantum yield
φ	quantum yield
r	distance between thermalized pairs or molecules
$f(r, \theta)$	the part of the dissotiated thermalized pairs in Onsager theory
$g(r, \theta)$	spatial distribution of the pairs
ΔE_{FP}	Pool-Frenkel decrease of the potential barrier
β_{FP}	Pool-Frenkel coefficient
v_{jk}	interaction energy between j and k states
v_c	critical value of the energy
w	statistical interval of the state energy
CTC	charge transfer complex
PVC	polyvinylcarbazole
PEPC	poly-epoxy-vinylcarbazole
m	carbazole group
m*	exited carbazole group
e*	eximer
k	constant rate of eximer disintegration

$\tilde{m}^*$	exited CT level
t	impurity level
t^*	exited exiplex
$\tilde{t}^*$	coupled ion-radical pair
E_{CT}	energy of the CT state
I_d	donor ionization potential
A_A	electron affinity
E_c	energy of coulombic interaction
E_r	resonance energy of donor-acceptor interaction
A	acceptor
D	donor
A^-, D^+	ions
cis $(CH)_x$	*cis*-polyacetylene
trans $(CH)_x$	*trans*-polyacetylene
1D	one-dimensional system
3D	three-dimentional systems
S^+-S^-	charged soliton-antisoliton pair
P^+-P^-	polaron pair
e-h-	electron-hole pair
PAC	polyphenylacetylenide of copper
lexan	poly (biphenol-A carbonate)
TPA	triphenylamine
IPC	isopropylcarbazole
R_o	radius of the local center
ρ_o	distance between local centers
j	localization parameter
SLM	spatiotemporal light modulator
TNF	2,4,7-trinitrofluorenone
ΔE	activation energy
T_{ef}	effective temperature
T_o	experimental temperature which characterize the transport system
e.m.f.	electromotive force

1 Introduction

Photoconductivity is one of the most informative phenomena in semiconductor physics. The light beam enables us to exite definite energetic levels strongly and obtain information about photogeneration, recombination and transport processes of the charge carriers. The history of photoconductivity shows the every essential step in research and application has been linked with the appearance of new photosensitive materials with specific physico-chemical properties.

Since the middle of this century, great success has been obtained in macromolecule synthesis and polymers have found broad application. The first papers on photoconductive polymers appeared in the early 1960s. The systematic research in this field came later [1–14]. A renewal of interest is especially evident in the last few years. The main reasons are the needs in the new materials for electrophotography, photothermoplastic recording, optoelectronics, holography, laser and integral optics. In addition polymers are the most disordered systems which are undergoing intense investigation now. Polymers presents unique opportunities for understanding the links between local states and chemical structure, the chemical and spectral sensitization mechanisms, the role of the excited states in various biological processes and so on.

Photoconductivity of polymers will be reviewed within the framework of semiconductor physics. The focus of attention will be the photogeneration and transport of the charge carriers, the relation between chemical structure and photoelectrical properties, and sensitization processes involving dyes and dopant molecules.

Polymers with saturated bonds, heteroatoms, heterostructures and polyconjugated ones are available now as photosensitive materials. Really one cannot expect a single mechanism to be reponsible for photoconductivity in so many diverse systems. However, there are a lot of verified factors which permit us to explain the main features of the photoconductive processes in polymers. The status and prospects of the application of polymeric photoconductors as prospective new electronic materials will be also analyzed for various types of photoconductors.

2 Photoconductivity and Related Phenomena

2.1 Basic Characteristics and Processes

Photoconductivity i.e. the change of the conductivity under the action of light $\Delta\sigma$ can be described by the expression

$$\Delta\sigma = e(\Delta\mu_n\Delta n + \Delta\mu_p\Delta p) \tag{1}$$

where e is the charge of the electron, Δn, Δp – concentration of the excess electrons and holes, $\Delta\mu_n$, $\Delta\mu_p$ – the change of the electron and hole mobilities.

The change of the electron concentration under the poorly absorbed light

$$\frac{\partial \Delta n}{\partial t} = \left(\frac{\partial \Delta n}{\partial t}\right)_{gen} + \left(\frac{\partial \Delta n}{\partial t}\right)_{rec} \tag{2}$$

where $\left(\dfrac{\partial \Delta n}{\partial t}\right)_{gen}$ defines the generation and $\left(\dfrac{\partial \Delta n}{\partial t}\right)_{rec} = (1/\tau)\Delta n$ recombination rate accordingly, τ the mean life time of the charge carriers.

The photocurrent i_{ph} can be written as

$$i_{ph} = eF(\tau/t_{tr}) = eFG \tag{3}$$

where t_{tr} – transit time of the charges between electrodes, F – generation velocity of the charge carriers, G – gain coefficient

$$t_{tr} = L/v_d = L/E\mu = L^2/V\mu \tag{4}$$

where v_d – drift velocity of the charges, E – electric field strength, μ – mobility of the free charges, L – distance between electrodes, V – voltage.

Taking into account 3 and 4 one can write

$$i_{ph} = e(F\mu\tau/L^2)V \tag{5}$$

It can be clearly seen that the photocurrent has to be linear function of the voltage, if μ and τ do not depend on the voltage. Really there are some mechanisms when this linearity does not hold. For example, if the voltage reaches the value, obeyed by the condition

$$V \cdot C = F\tau e \tag{6}$$

the external field introduces the volume charge into the photoconductor equal to the charge generated by the light. Here C is the specific capacity of the sample.

If $V \geqq F\tau e/C$ the transition times become equal to the time of the dielectric relaxation τ_r of the photoconductor and both times decrease with the voltage increase. Then Eq. (3) will become

$$i_{ph} = eF(\tau/\tau_r) \text{ or } (i_{ph}/eF)(1/\tau) = 1/\tau_r \tag{7}$$

The width of the photoconductor band transmission (ΔB), corresponding to the time constant of the photoconductor response, can be written $\Delta B = 1/2\pi\tau$

Then $G\Delta B = 1/2\pi\tau_r$ $\tag{8}$

where $G\Delta B$ is the product of amplification factor G and width of the band transition ΔB of the photoconductor, regarded as an amplifier in condition of the volume charge limited current. In such a case τ_r is equal to time constant of the photoconductor response.

In highly resistive photoconductors $1/2\pi\tau_r$, at low light intensity, may be less than unity. This means that photoconductors with a high gain coefficient will have a very slow response time.

In fact, Eq. (8) is fulfilled for the case of the free electrons without taking into account the trap levels. These levels exist in any real system. Response time of the photoconductor (τ_o) in such conditions will increase

$$\tau_o = (n_t/n)\tau \tag{9}$$

where n_t and n are the concentrations of the trapped and free electrons respectively.

All formulas mentioned above are fulfilled for any phenomenological photoconduction model. Microscopic description needs detailed research of the quantum yield, parameters of the generation, recombination and transport processes.

2.2 Experimental Methods

Identical methods for investigation of photoconductivity can be used for inorganic and organic semiconductors. Polymer semiconductors as a rule have very high resistance. For such materials the main information about the photoconductive mechanisms and properties may be obtained by two methods: electrophotographic (or discharge method) and time of flight (or transit method). Both methods are successfully applied for materials with low mobilities, less than $10^{-4}\,\text{m}^2\,\text{V}^{-1}\,\text{s}^{-1}$, which are the usual values for polymer semiconductors.

The main schemes of the discharge (a) and transit time (b) methods are shown in Fig. 1.

The sample on the conducting plate is charged by the corona discharge up to the surface potential V_o in the first method. Under the action of the strongly absorbed light one can observe the change of the surface potential due to the discharge current.

The discharge velocity at $t = 0$

$$\left(\frac{dV}{dt}\right)_o = \phi eI/C \tag{10}$$

where $\phi-$ generation efficiency of the charge carriers
    I – light intensity
    C – sample capacity

So the derivative dV/dt at the initial stage gives the direct measurement of the quantum efficiency of the generation.

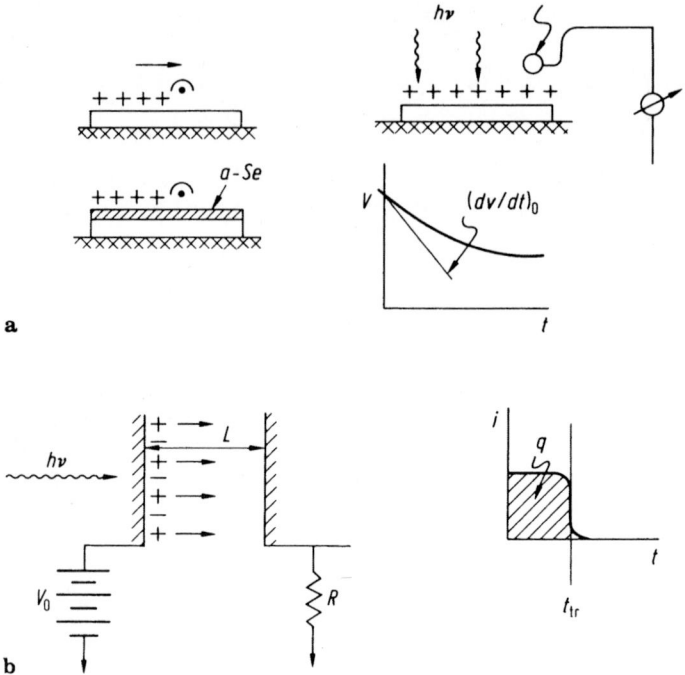

Fig. 1a, b. Typical schemes of the electrophotographic (**a**) and time of flight (**b**) methods for photoconductivity measurements in polymers [11]

The time of flight method permits us to investigate transport processes in highly resistive systems. The sample is inserted between two transparent conducting electrodes and controlled by constant voltage. If we illuminate the front electrode by strongly absorbed light, the excited excess carriers drift in the electric field. Taking into account the polarity of the electric field one can define the sign of the dominant charge carrier.

For positive lit electrodes one can register the drift of holes, and for negative ones- the drift of the electrons. The photosensitizer (for example Se) may be used for carrier photoinjection in the polymer materials if the polymer has poor photosensitivity itself. The analysis of the electrical pulse shape permits direct measurement of the effective drift mobility and photogeneration efficiency. The transit time is defined when the carriers reach the opposite electrode and the photocurrent becomes zero. The condition $RC \ll t_{tr}$ and $\tau_r \gg t_{tr}$ should be obeyed for correct transit time measurement. Here R – the load resistance, τ_r – dielectric relaxation time. Usually $t_{tr} \approx 0$, 1–100 ms, $RC < 0.1$ ms and $\tau_r > 1$ s. Effective drift mobility may be calculated from Eq. (4). The quantum yield (photogenerated charge carriers per absorbed photon) may be obtained from the photocurrent pulse shape analysis.

It was also shown that the transit time method permits us to obtain important and new information about statistical regularities of the charge transport in disordered systems [15]. It is so called dispersive transport. The definition of the transit time is not so simple, especially when we deal with dispersive transport. The transition time in this case can be obtained from photocurrent versus time dependence on a double logarithmic scale.

2.3 Models of Charge Generation and Transport

A lot of energetic models are applied for understanding of the photoconductive properties of the polymers. It is clear that as a rule polymers are heterogeneous or amorphous materials, so the band model meets with some difficulties. The weak point is an explanation of the low values of the drift mobilities which are much less than $10^{-4}\,m^2\,V^{-1}\,s^{-1}$ and can be of the order $10^{-10}-10^{-12}\,m^2\,V^{-1}\,s^{-1}$. So the length of the free path of the charge carriers becomes less than the size of the intermolecule link which is patently nonsense. So one can conclude the important role of the excitonic processes. The excitonic model considers that the initial act of the light absorption leads to creation of the neutral excited state – exciton. The latter may dissociate with generation of the free charge carriers due to the interaction with the interface, surface, impurities, other excitons and so on. The localized Frenkel type excitons are found in organic molecular materials.

Charge transfer states (CT) are often found in molecular systems side by side with excitonic states. CT states describe polar nonconducting states bound by coulomb interaction of the electron-hole pairs. CT states may be ionized with localization of the charges on definite molecules.

Two-step photogeneration mechanisms of the free charge carriers in polymers are considered frequently in the frame of the germinate recombination theory developed by Onsager [16]. At the first stage the photons generate bound electron-hole pairs. The part of the absorbed photons which generate the thermalized pairs gives the initial quantum yield φ_0, which does not depend on the electrical field. The second stage is the dissociation of the bound pairs in the combined action of the coulomb interaction and applied electric field. The part of the thermalized pairs separated by the distance r under the angle θ to the applied electric field, which dissociate – $f(r, \theta)$ obeys the Onsager equation

$$f(r, \theta) = \exp(-A) \exp(-B) \sum_{m=0}^{\infty} \sum_{n=0}^{\infty} \frac{A^m}{m!} \frac{B^{m+n}}{(m+n)!} \tag{11}$$

Here $A = 2q/r$; $B = \beta r(1 + \cos\theta)$; $q = e^2/2\varepsilon kT$, $\beta = eE/2kT$, ε – dielectric permitivity, e – the electron charge, kT – thermal energy.

The common efficiency of the photogeneration (quantum yield)

$$\varphi(E) = \varphi_0 \int f(r, \theta, E)\, g(r, \theta)\, d\tau \tag{12}$$

where $g(r, \theta)$ defines the spatial distribution of the pairs, $d\tau$ – the element of the volume in spherical coordinates.

So the free charge carrier generation depends on the quantum yield of the creation of the thermalized pairs and the probability of their dissociation.

Usually, one considers isotropic spatial distribution of the pairs and equal separation distance r for all pairs. Then

$$g = (4\pi r_o^2)^{-1} \delta(r - r_o)$$

and common solution follows the formula

$$\varphi(r_o, E) = \varphi_o\left[1 - \left(\frac{kT}{eEr_o}\right)\sum_{j=0}^{\infty} I_j\left(\frac{e^2}{\varepsilon kTr_o}\right)I_j\left(\frac{eEr_o}{kT}\right)\right] \tag{13}$$

where $I_j(x)$ – recurrence formula.

At low electric fields ($E < 3 \times 10^4$ V cm^{-1}) the theory gives

$$\varphi(E, T) = \varphi_o\left(1 + \frac{e^3E}{2\varepsilon(kT)^2}\right)\exp(-r/r_o) \tag{14}$$

The usual analysis of the experimental data consists of the measurement of the quantum efficiency versus electrical field and comparing the results with results theoretically calculated for different values of the parameters r_o, ε, T.

Gaidelis made critical analysis of Onsager's model and proposed the hopping scheme of the charge carrier photogeneration [17, 18]. Hopping and tunnel models for charge transfer are often used for explanation of the transport properties in organic polymeric photoconductors [4–14].

The excited π-electron may tunnel through a potential barrier in the free state of the neighbouring molecule preserving the energy. The probability for tunnel transition is as a rule, more than the probability of the returning to the initial state. Apparently the energy of the potential barrier may be considered equal to the molecule ionization potential. The barrier form depends on the coulomb potential between the electron and positive ion and affinity of the neutral molecule.

In the hopping model the charge moves from molecule to molecule over the barrier. The transfer process has to be thermally activated as in another disordered systems. The velocity of the charge carriers depends on the phonon velocity. Two steps are essential in the transfer of the charge carriers in the polymers. Intramolecular transfer may be characterized by the low activation energy, high conductivity and may appear during the alternating current measurement. Intermolecular transfer depends on the form and size of the intermolecular barriers and imperfection of the supermolecular structure.

The release of the trapped carriers may be stimulated by the electric field. In the Pool-Frenkel model the decrease of the potential barrier of the trapped electron by the electric field (ΔE_{Fp}) obeys the formula

$$\Delta E_{Fp} = (eE/\pi\varepsilon)^{1/2} = \beta_{Fp}E^{1/2} \tag{15}$$

where E – the electric field strength, β_{Fp} – Pool-Frenkel coefficient. This effect is due to coulomb interaction between released electrons and positive charges and strongly appears when the conductivity is limited by the volume charge.

Doping of the main polymer matrix by various types of donor or acceptor molecules including dyes, is a natural way of increasing the photosensitivity. One may observe the sharp change of the system's properties beginning with a definite concentration of the dopant molecules. The explanation of the phenomena may be obtained taking into account Anderson's model [14]. This model gives the answer as to which lattice disorder the charge transport become localized.

The conditions for realization of the localized and delocalized states are

$$\langle V_{jk} \rangle < V_c = \alpha^{-1}W - \text{localized state}$$

$$\langle V_{jk} \rangle > V_c = \alpha^{-1}W - \text{delocalized state}$$

$$\alpha = 6 \div 28$$

where V_{ik} – the interaction energy between j and k states; W – statistical interval of the state energy, V_c – critical value of the energy which is of the order W.

Localized states may exist in the case of the diagonal disorder but are impossible with undiagonal disorder. This model is used for analysis of the exciton transfer in various organic media. The model does not take into account the interactions with phonons and needs the existence of impurity bands. The last one is not so evident for organic materials.

The percolation model, which can be applied to any disordered system, is used for an explanation of the charge transfer in semiconductors with various potential barriers [4, 14]. The percolation threshold is realized when the minimum molar concentration of the other phase is sufficient for the creation of an infinite impurity cluster. The classical percolation model deals with the percolation ways and is not concerned with the lifetime of the carriers. In real systems the lifetime defines the charge transfer distance and maximum value of the possible jumps. Dynamic percolation theory deals with such case. The nonlinear percolation model can be applied when the statistical disorder of the system leads to the dependence of the system's parameters on the electrical field strength.

We mentioned the main models for generation, transfer, and recombination of the charge carriers in polymers. Very often, these models are interwoven. For example, the photogeneration can be considered in the frame of the exciton model and transport in the frame of the hopping one. The concrete nature of the impurity centers, deep and shallow traps, intermediate neutral and charged states are specific for different types of polymers. We will try to take into account these perculiarities for different classes of the macro-molecules materials in the next sections.

In addition to photoconductivity, there are a lot of photovoltaic phenomena observed in polymer photoconductors [14]. The most famous ones are the photo-emf at the Schottky barrier due to the separation of the electron-hole pairs in the electrical field at the photoconductor electrode interface; photo-emf at the interface between p and n regions of the semiconductor; Dember photo-emf due to the different mobilities of the electrons and holes.

2.4 Sensitization of the Photoconductivity

Various types of the photoconductive polymers are available now. The photoconductivity of such materials may be essentially increased by means of the chemical and spectral sensitization [12–14]. Spectral sensitization is connected with the appearance of the photosensitivity in the new spectral bands and the chemical sensitization with the increase of the proper sensitivity. As a rule both types of sensitisation may take place in the photoconductor at the same time. The first data about chemical and spectral sensitization in organic photoconductors appeared in [19, 20]. The example of the chemical and spectral sensitization of the photoconductivity by dyes in polymeric copper-phenylacetylenide is presented in Fig. 2. Later on it was proposed that not only low molecular weight compounds but polyconjugated polymers could also be used as sensitizers [21] having broad absorption bands and high thermostability compared with dyes. Now it is clear that various types of molecules may be used as a photosensitizers.

The effectiveness of the spectral sensitization depends on many factors. As a rule, a spectral sensitization process needs the thermal activation energy of the

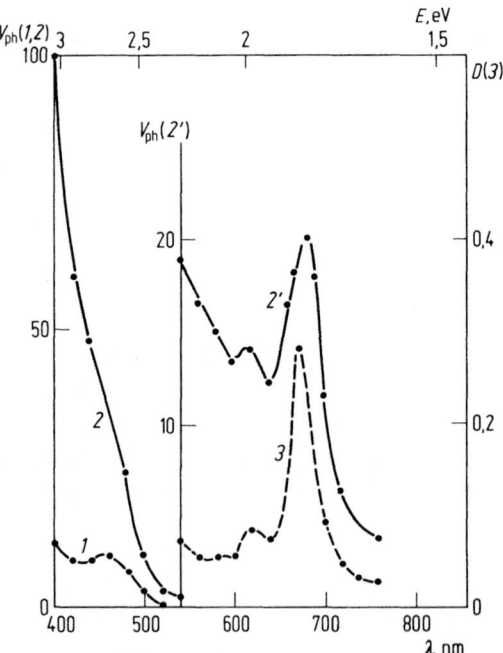

Fig. 2. Intrinsic (*1*), chemical (*2*) and spectral (*2'*) photo e.m.f. of polycopperphenylacetylenide sensitized by chlorophyl. (*3*), absorption spectrum of the sensitizer in solution [19]

order 0.05–0.1 eV. Optimum sensitization occurs when the dye molecule concentration is such as to cover 30–40% of the photoconductor surface. For polymer heterogeneous systems the weight concentration of the sensitizer has to be of the order 1%.

Two main models are usually discussed for the mechanism of the spectral sensitization. The excitation of the sensitizer by absorbed light and electron transfer from the excited sensitizer to the semiconductor is the first model. The alternative mechanism consists of the transfer of the excitation energy from the sensitizer to the semiconductor. This energy is used for photogeneration of the charge carriers in the sensitized photoconductor. In the first case the excited singlet level of the sensitizers has to be located above the conduction band of the semiconductor for realization of the electron transfer. For hole transfer the basic sensitizer level has to be located lower than the valence band of the sensitized photoconductor. The energy transfer mechanism does not need a special mutual location of the semiconductor and sensitizer levels.

Most of the modern theories of the photoconductivity sensitization consider that local electron levels play the decisive role in filling up the energy deficit. The photogeneration of the charge carriers from these local levels is an essential part of the energy transfer model. Regeneration of the ionized sensitizer molecule due to the use of the carriers on the local levels takes place in the electron transfer model. The existence of the local levels have now been proved for practically all sensitized photoconductors. The nature of these levels has to be established in any particular material. A photosensitivity of up to 1400 nm may be obtained for the known polymer semiconductors. There are a lot of sensitization models for different types of photoconductors and these will be examined in the corresponding sections.

3 Polymers with Saturated Bonds in the Backbone Chain

The following groups within the broad classes of the polymers with saturated bonds in the backbone chain may be distinguished for photoconductive polymers: polymers with developed π-conjugated side chromophores; charge-transfer complexes (CTC); polymers and complexes sensitized by dyes; polymers without conjugated chromophore groups. The light absorption in such systems as a rule leads to the excitation of the chromophore groups with a consequent charge or energy transfer to the unexcited sites of the molecules. The most important group is poly N-vinyl carbazole and its complexes. Most saturated polymers absorb in the UV region and the sensitization by dyes or CTC formation is necessary for photoconductivity optimization in the visual spectra. In all the above mentioned polymers, light absorption has a molecule nature because of the small intermolecular forces. The weak interaction between molecules leads to narrow bands and low mobilities of the charge carriers. So these materials differ from inorganic mono-crystals or disordered semiconductors, and low molecular organic semiconductors.

3.1 Polyvinyl Carbazole and its Analogs

3.1.1 Photogeneration of Charge Carriers

The structural formulas of the poly N-vinyl carbozole (PVC) and poly epoxy-propyl carbazole (PEPC) are

$$-CH-CH_2-$$

(PVC) N

(PEPC)

$$-O-CH-CH_2-$$
$$CH_2$$
$$N$$

The photoconductivity in PVC was observed for the first time by Högl [22]. It was shown that even insignificant traces of the donor-acceptor molecules had a striking effect on the photosensitivity. As it was demonstrated later, the donor-acceptor interactions play the decisive role in photoelectrical processes in polymers. The main analog of PVC is PEPC which has better mechanical properties and higher photosensitivity [17, 18]. Optical properties of PEPC are very close to PVC due to the light absorption by the analog chromophore groups.

Intrinsic absorption spectra of the carbazole polymers are in the UV range, but photosensitivity extends up to 1 micron. There are different mechanisms for the charge generation in every spectral range. In visual and near-IR spectra, photogeneration takes place due to the photoionization of impurities or traps, occupied in the dark. Exiplex formation with electron-donor groups or molecules happens in the intrinsic region. The exiplex appears to be a nonstationary donor-acceptor complex between excited donor molecule and unexcited acceptor molecule. Exiplex dissociation takes place spontaneously or may be increased by the electric field. Direct molecule ionization at wavelengths of less than 200 nm is the main process of charge carrier photogeneration.

A detailed analysis of photogeneration in the frame of Onsager's theory was reported in Ref. [23]. The calculated dependence of the relative quantum yield of the free charge carriers photogeneration versus electric field strength for different thermalization distances r_o at room temperature and dielectric permitivity equal to 3 is shown in Fig. 3 [11, p. 248]. Experimentally the quantum efficiency of the photogeneration increases step by step with increasing excitation energy (Fig. 4). The thermalization distances τ_o were 3.0; 2.8 and 2.6 nm for the third, second and first excited states respectively. The quantum efficiency dependences versus

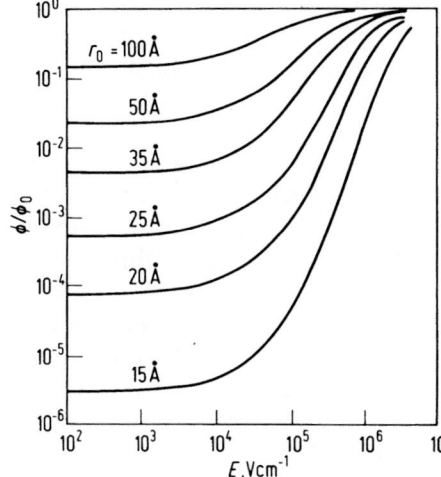

Fig. 3. Escape probability predicated by Onsager theory for isotropic initial distribution of electron-hole pairs. T = 296 K; static permittivity of $\varepsilon = 3$; thermalization distance of r_0 [11, p 248]

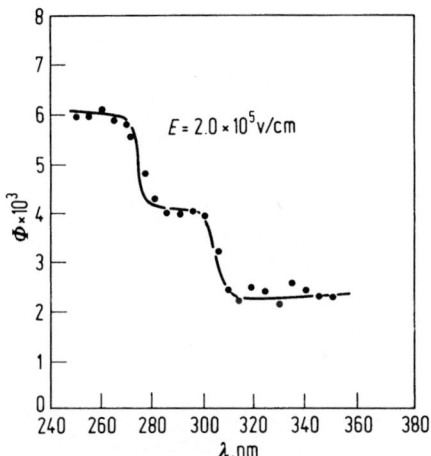

Fig. 4. Quantum efficiency of carrier generation in PVC versus excitation wavelength [23]

temperature vary for different thermalization distances. The temperature influence on the quantum efficiency at constant r_0 is stronger for low electric fields. It may be explained by the fact that mobility and diffusion coefficients have an activation energy. The temperature rise promotes the charge carriers separation and decreases the part played by the electric field strength in this process. The experimental data correlate with the theoretical ones with the assumption that the initial quantum yield increases with the increase in the temperature and constant thermalization distance.

The explanation of the nonmonotonic increase in the quantum efficiency (Fig. 4) may be connected with the high concentration of the intermediate exiplex sites [4]. The nature of these sites may be connected with deep acceptor traps located at 1.0–1.3 eV above the valence band. The confirmation of the short lived ion-radical pairs in PVC was proved by the magnetic field action on the fluorescence and the photoconductivity [21–25]. This field increased photoconductivity of the initial and doped PVC by 5–6% at wavelengths below 400 nm and did not influence conductivity and photoconductivity at wavelengths above 400 nm. The positive magnetic effect was sensitive to the electric field strength, temperature, and dopant concentration. The same action was determined for exiplex fluorescence. So the exiplex states play the key role in initial and doped PVC. It was shown that exiplex and ion-radical pairs formation took place due to the interaction between excited chromophore groups and oxygen or photooxides.

Another photogeneration mechanism, shown in Fig. 5, was proposed by Pearson [4, p. 350]. The intrinsic photogeneration processes are located to the left of the dashed line in the figure. The light absorption by carbazole group m leads to the excited state m*, which can relax in lower electron excited state or by means of diffusion migrate to the site energetically profitable for eximer e* formation. Eximer disintegration with constant rate k_F leads to eximer fluorescence. It appeared that the eximer fluorescence life time did not depend on the electric field strength. This and another experimental facts permitted one to exclude the charge carrier generation via the eximer stage. The excited state m can dissociate on the carrier pairs. Recombination of such pairs results in excited CT level m̃* In this case one has to observe the dependence of the charged generation on the electric fields. This was proved by experiment with close correspondence to Onsager's theory.

The impurity generation route is shown in Fig. 5 to the right of the line. The interaction of the relaxed states formed from m* and unexcited impurity t results in exiplex t* formation. This exiplex can decay thermally or form coupled ion-radical pairs ĩ*, which may dissociate in the electric field. For explanation of the absorption and photoconductivity spectrum correlation it is necessary to assume a very high concentration of exiplex sites.

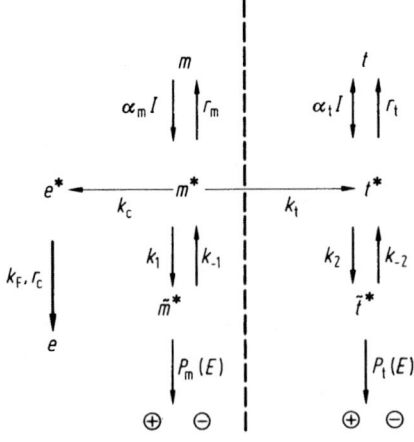

Fig. 5. Main photogeneration processes in PVC [4, p 350] α – absorption coefficient, k – dissociation constants, r – recombination constants, p – probability of the pair generation

The photogeneration mechanism was investigated by time of flight method [26, 27]. The discrepancy with Onsager's theory was remarked for weak and mean electric field strength. The photogeneration model with discrete intermediate states [17] gives the best description of the experimental results. The main route for charge carrier generation in air is the dissociation of the excited CT complexes formed by polymer chromophore groups and oxygen molecules as it was shown by the magnetic field method [24].

The main experimental results of charge carrier generation in PVC was explained in the frame of the Pool-Frenkel model [28–30]. The dependence of the recombination time on electric field was due to the change of the mobility in the electric field. Germinate recombination of the electron-hole pairs was investigated by means of luminescence decay characteristics [31].

3.1.2 Transport of Charge Carriers

Charge carrier transfer in carbazole polymers, as shown experimentally, is conditioned by impurity molecules which remain after synthesis or are deliberately introduced. Detailed analysis of the mobility was made by Huges [5, p. 158–189] and Gill [6, pp. 303–332]. The mobility of charge carriers in PVC is in the range from 10^{-7} to $10^{-12} \, m^2 \, V^{-1} \, s^{-1}$ and strongly depends on temperature and electric field strength. Exponential dependence of the mobility versus temperature as a rule takes place.

The filled upper and free lower orbitals in PVC belong to the carbazole groups. So one can consider PVC as a disordered organic medium. The side chromophore groups play the role of traps for hopping charge carrier transport.

It was actually shown by the time of flight method [32–34] that coulomb type traps control the drift mobility. The concentration of such traps is $10^{15} \, m^{-3}$. The real mobility (without traps) was estimated [35] to be of the order $5 \times 10^{-8} \, m^2 \, V^{-1} \, s^{-1}$ with a thermal activation energy of 0.28 eV. There are no correct data as yet confirming the impurity hopping model in PVC. The drift mobility is due rather to the jumps between neighbouring molecules and not shallow traps of the semiconductor.

Dispersive transport in PVC was investigated. The results of Pfister and Griffits obtained by the transit method are shown in Fig. 6. The hole current forms at temperatures > 400 K clearly show a bend corresponding to the transit time of the holes. At lower temperature the bend is not seen and transit time definition needs special methods. The pulse form shows the broad expansion during transition to the opposite electrodes. This expansion corresponds to the dispersive transport [15]. The super-linear dependence of the transit time versus sample thickness did not hold for pure PVC. This is in disagreement with the Scher-Montroll model. There are a lot of reasons for the discrepancy. One reason may be the influence of the system dimensions. It is quite possible that polymer chains define dimension limits on charge carrier transfer.

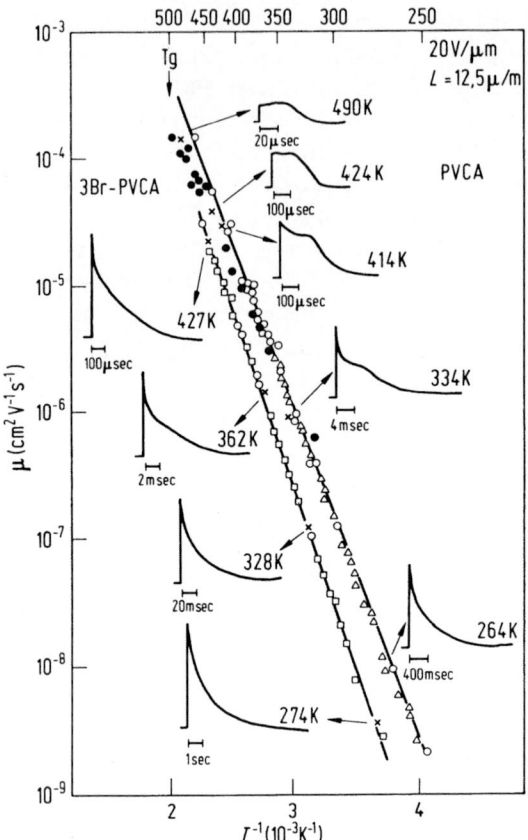

Fig. 6. Hole mobility versus temperature and pulse hole current forms in PVC and 3-Br-PVC [11, p 236]

The influence of the light intensity on the hole transfer was marked [36]. The good agreement between experimental and theoretical data for dispersive transport was observed only for a low intensity of the exciting light.

Special attention was paid to the influence of the morphology on the charge transport. Experimental data on the mobilities for different types of the carbazole polymers are shown in Fig. 7 according to Grifits et al. [11]. The disorder decrease leads to a decrease in the mobility and simultaneously to an increase in the dependence of the mobility on the electrical field strength. In amorphous polymers, we have significant ordering of the carbazole groups inside the chains and their parallel arrangement. The distance between carbazole groups in the chain is 0.41 nm and between chains 0.93 nm. So the last distance may restrict the charge transfer between chains. The substitution in position 2 by Br proved to have no great influence on inter and intra chain order. So the 2Br-PVC structure is analog to the PVC structure. The substitution by Br in position 3 gives quite another polymer structure leading to the absence of order. The polymer cannot be crystallized. The density of all polymers in spite of their different structures is nearly the same. One can see that the thermal activation

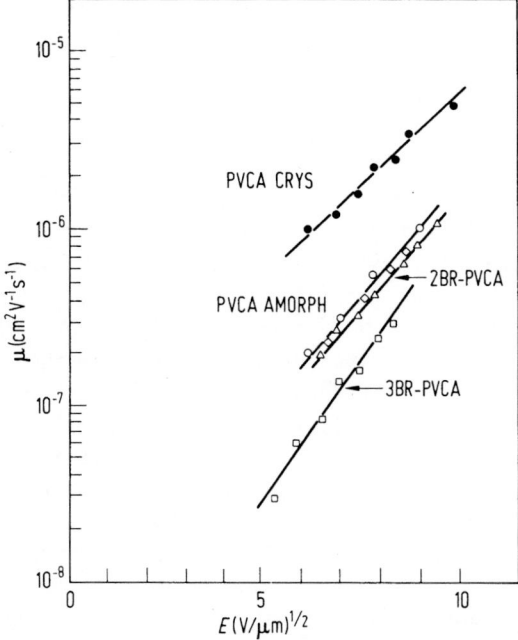

Fig. 7. Charge carrier (hole) mobility versus electric field strength in vinylcarbazole polymers [11, p 242]

energy of the mobility is larger in disordered polymers (Fig. 7). The influence of the sterical and morphological factors on the charge carrier mobility was examined in a lot of papers [37–42].

3.2 Donor-Acceptor Complexes of Carbazole-Containing Polymers

Intrinsically carbazole containing polymers are photosensitive in the UV range of spectra. The applications of such polymers in electrophotography and related processes need sensitization to the visual wavelengths. The most acceptable method is charge transfer formation between polymer donor and acceptor molecules. Hoegl pointed out that 0.1–2% of acceptor molecules inserted in the polymer matrix lead to a substantial increase in the photoconductivity especially in the CT bands. Subsequently, a lot of paper were published for CT-carbazole-containing complexes and such materials were used in photosensitive processes. Various types of molecules were used as a photosensitizers.

Sensitized photoconductivity appears in CT bands with energy E_{CT}

$$E_{CT} = I_D - I_A - E_C + E_R$$

where I_d – donor ionization potential, A_A – electron affinity of the acceptor, E_c – coulomb interaction energy between acceptor and donor ions, E_R – resonance energy of the interaction between $(D...A^-)$ and $(D...A)$. The equilibrium is shifted to the ion forms in the excited states.

High photosensitivity was obtained for CT of the PVC with 2,4,7-trinitro-fluorenone (TNF) [43, 44] which absorbs light up to 700 nm. This model system was used for optimization of the photoelectrical parameters for electrophotography and phototermoplastic recording. Various aspects of the charge generation and transfer for PVC-TNF complexes were considered by Pope and Swenberg, Huges, Mort, Gill, Pfister and other [3–14].

Xerographic photosensitivity spectra for CT complexes PVC-TNF and PVC-2,4,7,9-tetranitrofluorenone (TeNF) are presented in Fig. 8 [45]. It should be noted that xerographic spectra look like absorption spectra and thus similar explanations are to be expected for the photogeneration process. The latter can be divided into two steps. After the light absorption the excited state may be ionized or transferred to the ground state. As a rule the photogeneration processes in CT PVC complexes are analysed in the frame of Onsager's theory. Yokayama et al. proposed that absorption of a photon produces higher excited states which firstly deactivate to a lower excited state from which thermalization occurs [46–48]. From the values of dv/dt (Fig. 8) the quantum efficiency ϕ spectra were calculated. The results are presented in Fig. 9 [45]. The most interesting feature is that ϕ seems to be constant over a large range of the visible wavelengths. This clearly means that the charge separation process (thermalization) occurs from only one discrete energy level, corresponding to the absorption band. After the plateau ϕ decreases drastically. Approximately the same behavior was observed for the spectral dependence of the thermalization length r_0. For both acceptor r_0 is approximatively 3.2 nm. The results obtained confirm the previous models for the photogeneration process according to Onsager's theory.

However for the same CT complex PVC-TNF it was found that quantum efficiency had two different spectral dependence for light energy lower and above

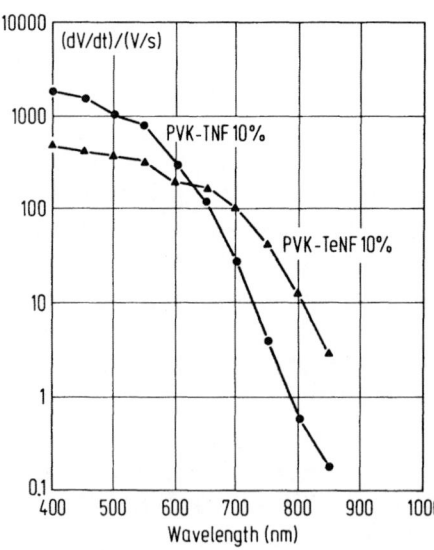

Fig. 8. Xerographic photosensitivity spectra for PVC-TNF [1] and PVC-TeNF CT complexes [45]

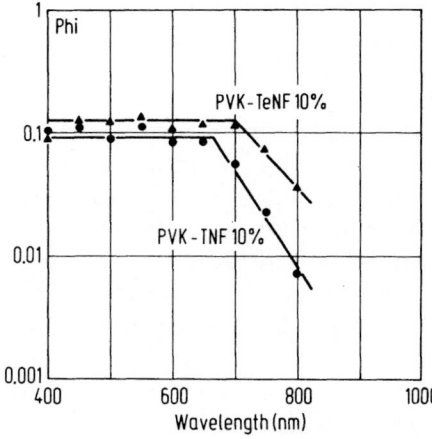

Fig. 9. Photogeneration quantum efficiency spectra for the two CT complexes [45]

2.7 eV [17, 18]. Impurity and exciton models were proposed for the photo-generation process [1, 7, 18, 28, 42]. The dependence of the quantum efficiency of the photogeneration on the electric field may be explained by the Pool-Frenkel model. The detailed analysis of the magnetic spin effects for chemical sensitization of the carbozole containing polymers was made in Ref. [49].

Carrier photogeneration processes versus polymer structure and various types of the dopant molecules were investigated by many authors [50–53].

Electron or hole nature of the charge transfer may be realized in PVC-TNF CT complexes depending on the concentration of the components [54]. The contribution of holes and electrons in conductivity is equal at the PVC-TNF ratio 1:1. Gill's experimental data for mobilities is shown in Fig. 10. The sharp concentration decrease of the neutral, noncoupled complex carbazole groups, leads to the fast decrease of the hole mobility. The electron mobility increase with increase of the TNF content was also accompanied by an increase of the

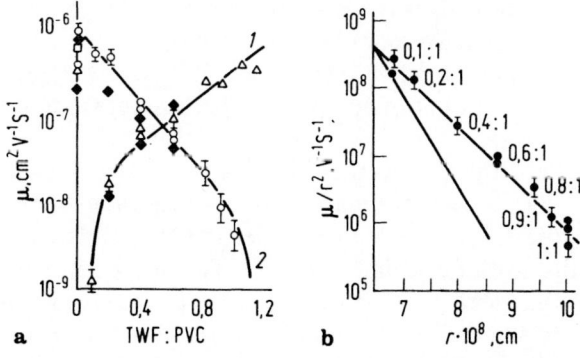

Fig. 10. Dependence of the charge carrier mobilities in PVC-TNF [54] **a** electrons (1) and holes (2) versus molar concentrations of the TNF-PVC at the electric field strength 10^5 V·cm^{-1}. **b** holes versus mean distance r between nonbound in complex carbazole groups

noncoupled groups in complex TNF molecules. The results obtained show that electron transfer originates via free TNF molecules and hole transfer via free carbazole groups. So the CT role reduces to the increase in the free charge carriers concentration but not to the drift facilitation.

At low acceptor concentrations, photoconductivity results from hole hopping between donor groups. The hole mobility versus the distance between noncomplexed carbazole groups obeys the formula $\mu \sim r^2 \exp(-kr)$ (Fig. 10b). This confirms the hole transfer by jumps between neutral carbazole groups. The lower line shows the hole mobility under conditions that all carbazole groups take part in the hole transfer. The details of the charge transfer may be derived from the temperature and field dependence of the mobility. The temperature dependence of the mobility obeys the law $\mu \sim \exp[-\Delta E/kT_{ef}]$ where T_{ef} is the effective temperature. Phenomenologically, the correlation $1/T_{ef} = 1/T - 1/T_o$ was established, where T_o – experimental temperature which characterizes the transport system. The materials with weak mobility dependence versus the electric field may be characterized by low T_o (high T_{ef}). So T_o is the important parameter for transport analysis in organic polymer systems.

The dependence of the drift mobility μ on the electric field is represented by formula $\mu \sim (\beta \cdot E^{1/2}/kT_{ef})$ which corresponds to the Pool-Frenkel effect. The good correspondence between experimental and theoretical quantity for Pool-Frenkel coefficient β was obtained. But in spite of this the interpretation of the drift mobility in the frame of the Coulombic traps may be wrong. The origin of the equal density of the positive and negative traps is not clear. The relative contribution of the intrinsic traps defined by the sample morphology is also not clear [17, 18]. This is very important in the case of dispersive transport. A detailed analysis of the polymer polarity morphology and nature of the dopant molecules on mobility was made by many authors [55–58].

3.3 Sensitized Photoconductivity in Carbazole Polymers

Chemical and spectral sensitization by dyes is a powerful method of increasing the photosensitivity. As a rule the dye plays a double role – the photogeneration sensitizer and desensitizer due to the change of the recombination and trapping processes. Various types of dyes with donor or acceptor properties were used for such purpose.

The photoconductivity and absorption spectra of PVC with pynacyanol as a sensitizer are given in Fig. 11 [59]. It can be seen that a photo response appears in the range of the absorption maximum of the monomolecular form of the dye. The results are typical for all sensitization data [60–62]. The addition of the dopant molecules leads to the change of photogeneration, transfer and recombination efficiency. The effectiveness of the sensitization increases with lowering of the first excited state of the dye molecule.

Magnetic field influence on the sensitized photoconductivity may explain the nature of the electronic processes. It was shown that PEPC photoconductivity

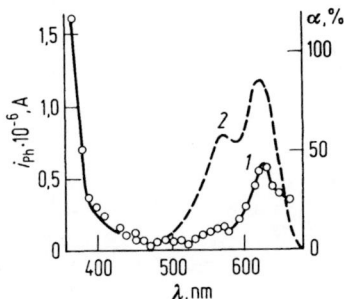

Fig. 11. Sensitized photoconductivity (*1*) and absorption (*2*) spectra of PVC with pinacyanol [59]

sensitization occurs only in the presence of the oxygen molecules [24, 49]. The influence of the magnetic field on the photocurrent and fluorescence is shown in Fig. 12. The magnetic field influence is due to the concentration change of the charge transfer complexes.

The shallow and deep levels play the important role in the sensitization process. Detailed research in this field has shown the presence of four local electron centers in the energetic spectrum of the sensitized PVC in the range 0.6–3.3 eV [63, 64]. The density of localized states was of the order 10^{18}–10^{19} cm^{-3}. These can play essential role in spectral and chemical sensitization due to their influence on photogeneration, recombination and charge transfer processes.

3.4 Heterogeneous and Multilayer Systems with Carbazole-Containing Polymers

Multilayer and heterogeneous systems allow us to increase our understanding of photogeneration and charge transfer processes and obtain the high photosensitivity needed for practical applications. Carrier injection into PVC of

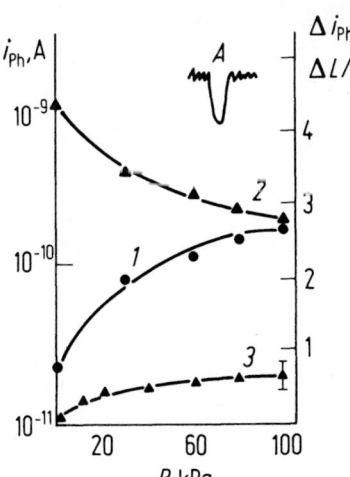

Fig. 12. Dependence of the photocurrent (*1*), magnetic effect on photocurrent (*2*) and fluorescence (*3*) in PEPC sensitized by Rhodamine on air pressure [24]

various types of the inorganic photoconductors – Se, Cds, CdTe [3–6, 11–14, 65–69] and organic ones [70–78] has often been used to increase photosensitivity.

Photosensitivity in a PVC-Se two layer system is equal to the Se photosensitivity in the strong absorption region [13]. The experimental data are presented in Fig. 13. Strong dependence of the photoresponse on the electric field strength was established. For n-Se and p-Se it was explained by the dependence of the quantum yield and charge carrier transfer from the electric field respectively.

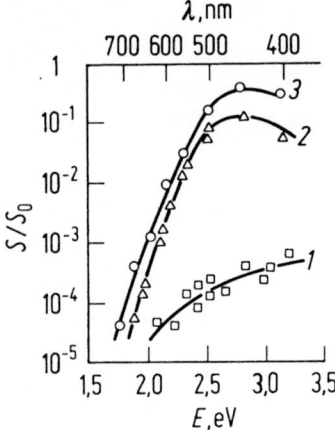

Fig. 13. Electrophotographic photosensitivity spectra for layers of PVC (*1*), PVC-*n*-Se (*2*), PVC-*p*-Se (*3*) [13]

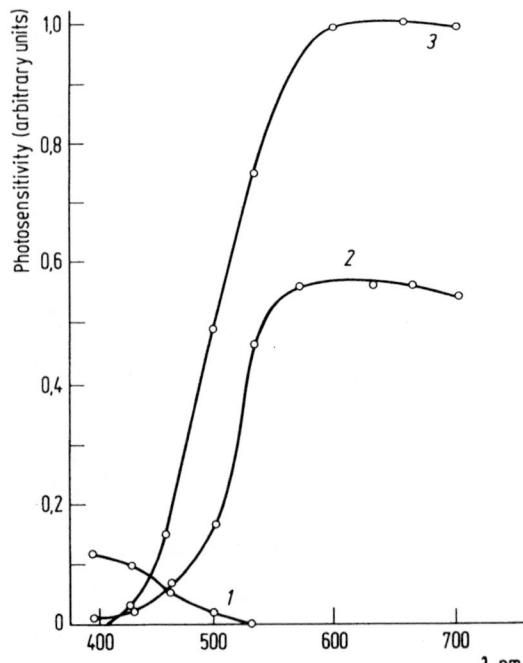

Fig. 14. Electrophotographic photosensitivity spectra for PVC (*1*), PVC-TNF (*2*), two layer PVC-PVC-TNF (*3*).[73]

Maximum photosensitivity was obtained for two layer system containing a thin PVC-TNF layer covered by thick PVC layer [73] (Fig. 14). Here CT complex realizers photogeneration function and PVC the transport one. Memory effects, p–n transitions, and others may be obtained for multilayers and polydispersed systems. The photosensitivity of the last ones strongly depend from the incapsulation velocity of the inorganic photoconductors in polymer matrixes [79–81].

3.5 Other Polymers and Their Complexes

Side by side with carbazole containing compounds other types of photoconducting polymers with saturated bonds in the backbone chain were investigated. Early references can be found in [7, 14]. These polymers have very high resistivity. Their possible usage in electrophotography was the main reason for studying their photoelectrical properties. Conditionally one can distinguish the polymers with large bulky side groups. The mechanism of the photogeneration and charge transfer in such material is analogous to the processes in PVC. The absorption and photosensitivity is situated usually in the near-UV region of the spectra. The compounds without side groups absorbs in vacuum ultraviolet and have photosensitivity in this region.

We may regard the commercial polymers which have broad technical application separately. The photosensitivity of such materials, as a rule, is caused by impurities and dopants. The main reason for studying their photoelectrical processes is the clearance of the stabilization problems.

Some types of the polymers were investigated in detail. The photoconductivity of polyethylene with quantum efficiency 10^{-5}–10^{-10} is caused by impurities, Schottky type contact injection, and hole transport [82, 83]. The crystallinity increase is accompanied by a photocurrent increase. There is no clear correlation between the chemical structure and the photocurrent.

Photoconductivity spectra for some vinyl polymers in the vacuum UV region is shown in Fig. 15 [13]. The close resemblance of the vacuum UV spectra for saturated and polyconjugated materials should be noted. The bathochromic shift is observed with the increase of the size of the side groups. As a rule the increase in the photosensitivity takes place with decreasing light wavelength. The mobilities of the charge carriers are in the range 10^{-8}–10^{-12} m^2 V^{-1} s^{-1}.

Photovoltaic volume effects have been investigated for the ferroelectric polymer–poly vinylidene fluoride [84, 85]. The photovoltage was of the order 4×10^4 V for open circuit. The addition of dyes shifts the photosensitivity to the longer wavelength.

Some research was carried out on biopolymer photoconductivity [14]. The significance of the charge carrier mobility up to 10^{-4} m^2 V^{-1} s^{-1} was pointed out.

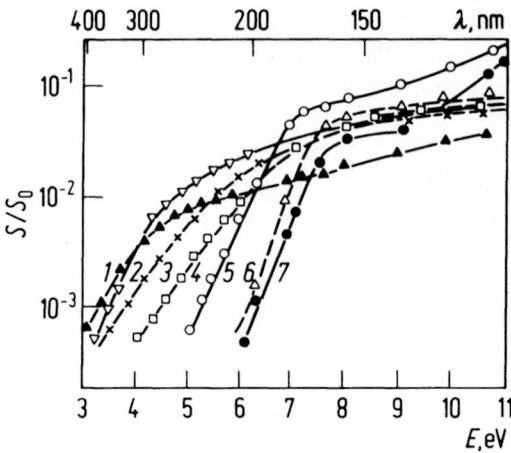

Fig. 15. Electrophotographic spectra for polymers: *1* – polydiphenylbutadiyne, *2* – polyvinylcarbazole, *3* – polyphenylacetylene, *4* – polyethylene, *5* – polyacrylonitrile, *6* – polyvinylchloride, *7* – polystyrene [13]

Photovoltaic and photoconductive phenomena for various types of CT complexes between saturated polymers and dopant molecules, heterojunctions between polymers and organic and inorganic photoconductors were also investigated in the last few years [86–92]. The quantum efficiency of the energy conversion of $10^{-3}\%$ was obtained for such systems and output power density of 3×10^2 mV cm^{-2}. The mobilities of the heterogeneous polymer systems with despersed inorganic photoconductors reach the value -10^{-3}–10^{-4} m^2 V^{-1} s^{-1}.

4 Polymers with Conjugated Bonds, Heteroatoms and Heterocycles in the Backbone Chain

The first papers concerning the photoconductivity of the polyconjugated compounds were published in the early 1960s [93–99]. At first, the research was conducted on powders and pressed tablets. It made the interpretation of the results difficult. New era has begun after synthesis of the polyacetylene by Shirakawa et al. [100]. An intense study of the electronic properties of polyacetylene was conducted by Heeger et al. and others [101–112]. Due to this research it became clear that the conjugation chain defects – solitons, polarons and bipolarons may be the charge and energy carriers in polyacetylenes. This conception stimulated a lot of research into the photoexcited states in polymers and became one of the most prominent features of polymer semiconductor physics. Polyacetylene is the ideal model system of the polyconjugated polymers.

Another important step was made by Wegner who obtained crystalline polydiacetylenes of high molecular weight exhibiting a fully conjugated backbone [113]. The theory describes the electron structure of the polydiacetylenes in the frame of the 1D wideband semiconductor. The polydiacetylene systems are simpler than those of polyacetylenes because the latter cannot be obtained as monocrystals. Polydiacetylenes are prospective materials for electronic devices because of the high mobilities of the charge carriers and interesting nonlinear optical properties.

Now the research in the photoconductive properties of the polyconjugated materials is growing fast. Heterocycle or heteroatom-containing polymers are involved in this process due to their excellent mechanical and electric properties. The sensitized photoeffect in polyconjugated materials was first observed in 1964 [19, 20] and the high significance of the increase in the photosensitivity of these compounds became apparent.

The main differences of polyconjugated materials from saturated polymers are the delocalization of the π-electrons along the macromolecules. One can expect bands which are sufficiently broad in the polyconjugated compounds, analog to the inorganic semi conductors in the condition of high long range order. Anisotropy of the electrical properties may be realized at high ordering of the macromolecule chains. Various types of defects make such polymers one of the most disordered systems which are being currently studied.

The nature of the photoexcited states is the essential key to understanding the photoconduction process. The initial step depends on the light absorption character. The molecular structure defines such absorption in polymers unlike the inorganic semiconductors. In the latter, there are broad bands inherent to the crystal lattice. The absorption bands are shifted in the long wavelength region of spectrum compared with the absorption of the isolated atoms and molecules. The close resemblance between absorption spectra of the solvents and solid states shows insignificant intermolecular interaction in the polymers. So the molecules remain individual in the solid state. The light absorption leads to the excited states, excitons, which dissociate with the creation of the free charge carriers. For every type of polymer, these states have their own nature and peculiarities. However, during the life time of the excited states the excitons may travel up to one hundred of the monomer links as was shown by luminescence methods.

Commonly, the increase in the conjugation chain length leads to the bathochromic shift of the absorption bands. The optical and photoelectrical properties analysis needs to take into account not only formal polyconjugation following from the molecule chemical structure but effective conjugation defined by sterical factors too.

There are no significant bathochromic shifts in the absorption and fluorescence spectra beginning from some critical meaning with the increase of the conjugation. This shows that continuous conjugation does not stretch on the whole chain length but includes only small, usually from two to seven monomeric links. One can understand why the inclusion in the chain of various bridge groups or heteroatoms, which decrease the conjugation, does not lead to the disappearance of the semiconductor properties. So the optimal conjugation length in polymers which defines the semiconductor properties is not more than a few links since they occur in low molecular organic solids. Therefore the initial act after the light absorption in polymers is the exciton generation by the short section of the conjugated chain. So the common nature of the low and high molecular weight semiconductors may be seen.

As to the charge transfer mechanism, one can consider that the alteration of the high conductive (conjugated parts) and dielectric (disordered or amorphous structure) regions are typical for organic polymers. So for the realization of photoconductivity, the generated electron-hole pairs have to drift in the electric field. The coincidence of the conductivity activation energy with the longwave absorption band edge and the red edge of the photoconductivity obtained for many polymers shows that photogenerated free carriers do not differ from those thermally generated. So the drift mechanisms for them may be regarded as identical.

The numerous defects inherent in organic polymers creates the donor or acceptor impurity levels. The low drift mobilities of the order 10^{-7}–10^{-12} $m^2 V^{-1} s^{-1}$ lead to the paradoxical situation where the length of the free travel distance for the charge carrier becomes less than the size of the separate molecule links. So the hopping or activated models are the most acceptable ones for polymers in such circumstances.

One can see the fruitfulness of the hopping and band model synthesis for heterogeneous polymer structures. In this case we may consider the charge transfer inside the conjugated section of the chain in the frame of the band model and transitions between these parts of the chain as jumps or as an activated surmounting of the barriers.

4.1 Polyacetylenes

Polyacetylene attracts constant attention as an excellent simple model of the polyconjugated polymer on which the main optical and electrical properties can be verified. The possibility of achieving metallic conductivities by doping opens real perspectives of practical application of conducting polymers. The complication is the strong interaction with oxygen. The reproducibility of results strongly depends on the synthesis and measurement conditions.

Polyacetylene consists of CH-group chains which form a quasi one-dimensional (1D) lattice [104–107]. There are *cis* and *trans* forms of polyacetylene.

On the terminology of the band model σ-bonds form the completely filled low band, and π-bonds make the partially filled band, which defines the electronic properties of the polyacetylene.

Active research of the electronic properties began in the early 1970s, when it was shown that polyacetylene may be synthesized as a flexible film with arbitrary and specially oriented fibrils.

Two lower states of the *trans*-$(CH)_n$ are energetically degenerated as follows from symmetry conditions. Theory shows that electron excitation invariably includes the lattice distortion leading to polaron or soliton formations. If polarons have analogs in the three dimensional (3D) semiconductors, the solitons are nonlinear excited states inherent only to 1D systems. This excitation may travel as a solitary wave without dissipation of the energy. So the 1-D lattice defines the electronic properties of the polyacetylene and polyconjugated polymers.

Due to the alteration of the bonds, *trans*-$(CH)_n$ is a semiconductor with the forbidden gap 1.5 eV. So this material has a principal difference from traditional organic semiconductors containing poorly interacting molecules and from polymers with saturated bonds without π-electrons. So *trans*-polyacetylene is closer to the inorganic semiconductors. Unlike classic covalent semiconductors, polyacetylene may be doped after synthesis at room temperature due to the open morphology and weak interchain interaction. During doping, charge generation in the polymer may occur owing to the charge transfer. The transfer takes place from polymer to the acceptor (A) and the polymer chain acts as a polycation in the presence of A^-. The polymer chain acts as a polyanion for donor D in the presence of D^+. The ions A^- and D^+ are inserted between polymeric chains.

4.1.1 Soliton Models of Charge Generation and Transport

Photoconduction and absorption spectra of *trans*-polyacetylene are presented in Fig. 16 [104].

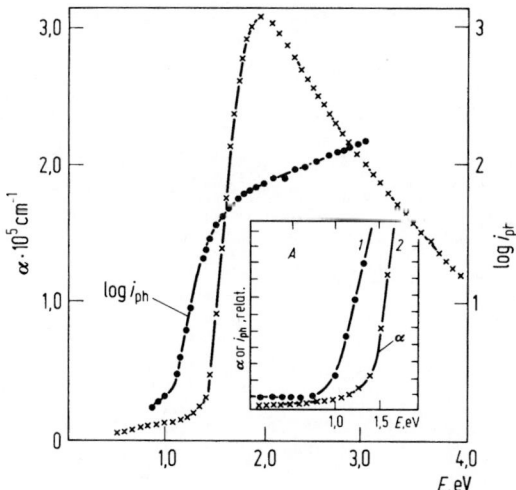

Fig. 16. Photoconductivity (*1*) and absorption (*2*) spectra of transpolyacetylene: *A* – photocurrent and absorption in linear scale [104]

The threshold of the photocurrent is at the energy of 1 eV. This value is less than the absorption edge energy, what proves the existence of deep levels inside the forbidden gap. The absence of the structure at the beginning of the interband transitions excludes the possibility that the photocurrent is due to the surface exciton dissociation. Attempts to observe good photoconduction in cis-$(CH)_n$ were unsuccessful. The photocurrent in cis-$(CH)_n$ was three orders of magnitude lower than in $trans$-$(CH)_n$.

These and other results on absorption and luminescence were the reason why the soliton model was proposed for the photoconduction process in $trans$-polyacetylene [103–110].

The main scheme is shown in Fig. 17. The photogenerated electron hole pairs transfer to the soliton-antisoliton pairs in 10^{-13} s. Two kinks appeared in the polymer structure, which separates the degenerated regions. Due to the degeneration, two charged solitons may move without energy dissipation in the electric field and cause the photoconductivity. The size of the soliton was defined as ~ 15 monomer links with the mass equal to the mass of the free electron. In the scheme in Fig. 17, the localized electron levels in the forbidden gap correspond to the free (+) and twice occupied (−) solitons. The theory shows the suppression of the interband transitions in the presence of the soliton. For cis-$(CH)_n$ the degeneration is absent, the soliton cannot be formed and photoconductivity practically does not exist.

A lot of research was carried out for proving existence of the soliton in $trans$-$(CH)_n$ by various methods - induced absorption and luminescence, electron spin resonance, transit photogeneration and so on. The references may be found in the monograph [14].

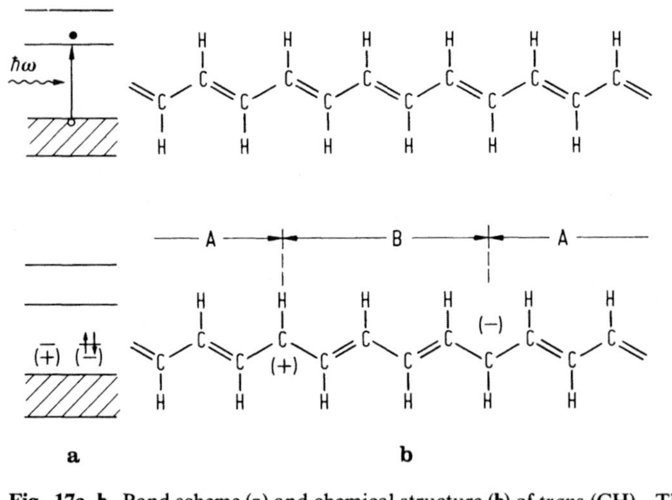

Fig. 17a, b. Band scheme (**a**) and chemical structure (**b**) of $trans$-$(CH)_n$. The polymer chain contains the charged soliton-antisoliton pair. In the middle of the forbidden gap there are levels connected with two solitons: not occupied (+) and twice occupied (−) [106]

The detailed scheme for relaxation of the photoexcited states and charge carrier generation (Fig. 18) was put forward on the basis of picosecond photo-induced absorption and bleaching [114]. The photoexitation relaxation may be divided into time intervals with different kinetics. At the first stage, the photoinduced electron hole pair is transformed to the charged soliton-antisoliton pair (S^+–S^-) in the chain. Electron hole pairs with a quantum yield of 10^{-2} are formed at different chains and are transformed to the polaron pairs (P^+–P^-). The most of the soliton-antisoliton pairs recombine in the ID chain after accidental floating with a mean time jump of the order 0.1–1 ps. The maximum soliton-antisoliton absorption is at 1,35 eV. Recombination of the polarons between the chains is the main process in an interval between nanoseconds and microseconds. Due to this process the charged soliton-antisoliton pairs appear. They exist for some milliseconds and form the free charge carriers causing photoconductivity. The absorption band at 0.5 eV is due to the formation of the charged solitons.

The soliton conductivity model for $trans$-$(CH)_n$ was put forward by Kivelson [115]. It was shown that at low temperature phonon assisted electron hopping between soliton-bound states may be the dominant conduction process in a lightly doped one - dimensional Peierls system such as polyacetylene. The presence of disorder, as represented by a spatially random distribution of charged dopant molecules causes the hopping conduction pathway to be essentially three dimensional. At the photoexitation stage, mainly neutral solitons have to be formed. These solitons maintain the soliton bands. The transport processes have to be hopping ones with a highly expressed dispersive

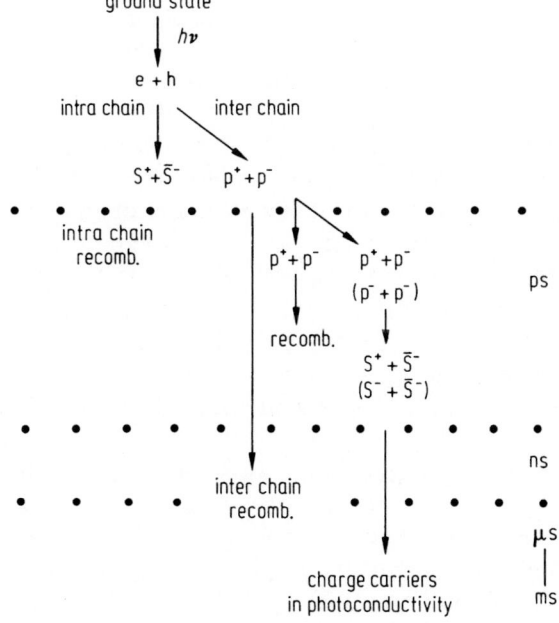

Fig. 18. A model for the relaxation mechanism of photoexcited $trans$-polyacetylene at different time intervals [114]

character. A great number of experimental results [101–103] confirm Kivelson's model.

The deficiency of this model is the assumption of only light doping. However the soliton conductivity has also been established in highly doped material and not only for polyacetylene, but for polyphenylene too, in which the solitons cannot exist.

The polaron-bipolaron model of conductivity was proposed for some polymers [116, 117]. Polaron in polyacetylene and n-phenylene is considered as a pair consisting of a free radical (neutral soliton) and an ion (charged soliton). The estimation of the link energy gives a value of 0.05 eV. With light doping, the localized polaron states appear in the forbidden gap. With high doping, localized polaron states form two polaron bands. For polymers without degenerated ground states, two charged defects cannot be separated and bipolaron formation takes place. Bipolarons are correlated pairs from charged soliton and antisoliton. It appears that in $trans$-$(CH)_n$ with low doping, polarons and solitons coexist and transfer the charge.

For polymers without degenerate ground states, the polarons are the only charge carriers. There are a lot of experimental and theoretical data confirming the formation of polaron and bipolaron bands in the forbidden gap of the polymer [101–107]. Frankevich confirmed the soliton and polaron participation in the charge transfer of the lightly doped polyacetylene, observing the conductivity change in a superhigh frequency magnetic field [118, 119]

4.1.2 Other Models of Charge Generation and Transport

One cannot say that the soliton model is the only one for the explanation of the optical and electrical properties of polyacetylene. There are a lot of defects in the polymer which can strongly influence the photoconductive processes. According to Kivelson's model, the time for hopping between the chains $\sim 10^{-7}$ s. But the lifetime of the carriers was shown to be less than 3×10^{-9} s [120]. So the mobile solitons, if they exist, have to be very fast intermediate states of the excitation relaxation in the chains. The non-solitonic nature of picosecond photoconductivity in $trans$-polyacetylene was strongly proved in [121–123] where much of the references concerned may be found. It was observed as being fast, with a relaxation time of 100 ps, and slow photoconductivity. The carriers are able to move freely along the polymer chain in the case of fast photoconductivity. The slow component consists of carriers persisting up to seconds with a time independent mobility, which is smaller by a factor of a hundred than that of the fast meaning of 10^{-4} $cm^2\,V^{-1}\,s^{-1}$.

The transport mechanism of the slow photoconductivity is determined by the hopping process. In the oriented samples it was observed as a strongly anisotropic mobility of the carriers with respect to the applied electric field $(\mu_{\parallel}/\mu_{\perp} = 50)$. The amplitude of the photoconductivity signal was also anisotropic with respect to the polarization of the incident light. For these reasons, photogenerated charges had to be attributed to carriers separated on different polymer chains (or chain-segments). Therefore they had to be polarons initially. Carriers remaining on the same chains recombine into neutral excitation.

Some difficulties of the soliton theory for explanation of the photovoltaic and other effects were also noted [124–127].

4.1.3 Nonlinear Elements with Polyacetylene

Nearly thirty papers have been published on the possible polyacetylene applications as nonlinear elements including rectifying cells [105, 128–134] and heterojunctions [135–139].

Schottky photovoltaic elements were proposed because of the good conformity of the solar spectrum with the polyacetylene forbidden gap. The photovoltaic spectra for $(CH)_n$-Al structure and volt-current characteristic for $(CH)_n$-nCdS heterojunction are presented in Fig. 19 and Fig. 20 respectively.

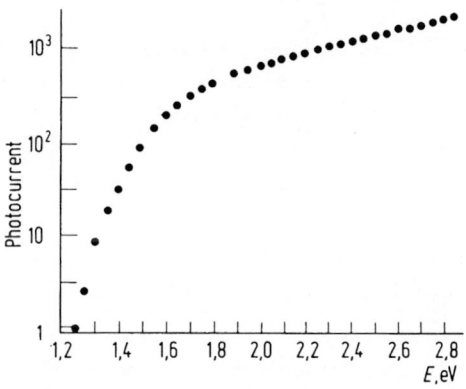

Fig. 19. Photovoltaic spectra for Schottky type barrier $(CH)_n$-Al [128]

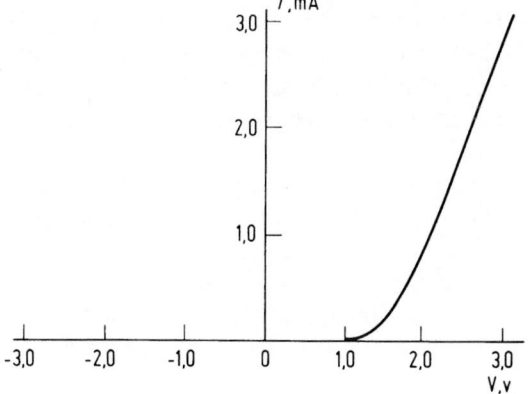

Fig. 20. Volt-current characteristic of heterojunction $(CH)_n$-n-CdS [136]

The energy conversion efficiency is estimated to be 0.3% relative to the energy absorbed within the depletion layer region. This conversion efficiency is remarkably higher than that of PVC-TNF. The rectification factor is 10^3, short circuit current 50 $\mu A \cdot cm^{-2}$, open circuit voltage 0.8 V, fill factor 0.25.

The heterojunctions of the polyacetylene were realized not only with inorganic photoconductors but also with organic polymers [139]. The results obtained show good similarity with barrier and heterojunction characteristics for inorganic semiconductors. Photoelectrochemical cell for solar energy conversion with polyacetylene electrodes and Na_2S, electrolyte had an efficiency of 1% at 2.4 eV [140]. The complicated phenomena take place at the electrode-electrolyte interface.

In conclusion, one can say that the possibility of a sharp change of the polyacetylene conductivity opens vast perspectives for new technologies in the field of molecular electronics. The cheap, simply made devices can be designed with small power consumption and broad excellent physico-chemical properties.

4.2 Polydiacetylenes

Crystalline polydiacetylenes, exhibiting a fully conjugated backbone, can be obtained [113] by thermal or photochemical polymerization of the monomer $R-C \equiv C-C \equiv C-R$, where R is the organic substituent. The polymeric structures of the type

$$\left(\begin{array}{c} R \\ \diagdown \\ C-C\equiv C-C \\ \diagup \qquad \diagdown \\ R \end{array} \right)_n \quad \text{and} \quad \left(\begin{array}{c} \diagdown \qquad R \\ C=C=C=C \\ \diagup \qquad \diagdown \\ R \end{array} \right)_n$$

were confirmed.

As it was pointed out earlier, the optical and photoelectrical properties of the polydiacetylenes may be explained in the framework of 1D crystal model in which the interaction between the electrons in the chain is much stronger than between the chains. The interaction energy differs by 100 fold. Considering the anisotropy of the optical and photoelectrical properties one can expect this fact. This was actually observed for monocrystals and films of the polydiacetylene. Absorption spectra of the monocrystal films are presented in Fig. 21 for varying light polarization [141].

Many authors have investigated the photoconductivity of the polydiacetylenes [142–171]. The main problem discussed concerns the nature of the initial act of the photoeffect. At first, most authors considered the exciton formation to occur at the beginning with consequent dissociation on the free carriers. Then it was shown the broad band existence for directions along the chains. The unification of the excitonic and band model of the free charge carrier generation was developed [146–150].

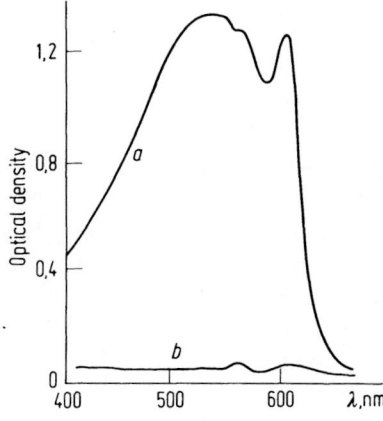

Fig. 21. Absorption spectra of the monocrystal poly-diacetylene film with $R = CH_3-C_6H_4-SO_3-(CH_2)_4-$ for the light polarized parallel (*a*) and perpendicular (*b*) to the macromolecule chain [141]

The photoconductivity and absorption spectra of the multilayer poly-diacetylene are shown in Fig. 22 [150]. The continuous and dotted line relate to the blue and red polymer forms respectively. Interpretation is given in terms of a valence to conduction band transition which is buried under the vibronic sidebands of the dominant exciton transition. The associated absorption coefficient follows a law which indicates either an indirect transition or a direct transition between non-parabolic bands. The gap energies are 2.5 eV and 2.6 eV for the two different forms. The transition is three dimensional indicating finite valence and conduction band dispersion in the direction perpendicular to the polymer chain.

An increase of the photocurrent at energies less than 2 eV was observed [151, 152] unlike the previous result. This was attributed to the localized impurity ionization up to 0.8 eV below the conduction band. The crystals are considered as model systems for the one and three-dimensional versions of Onsager's theory of germinate recombination.

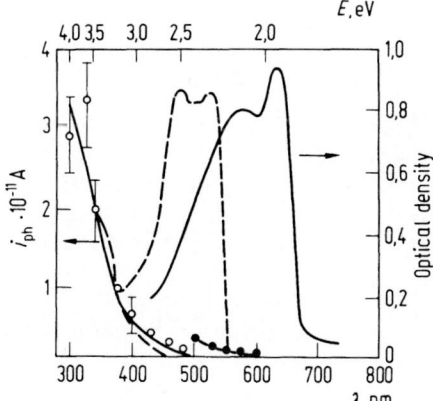

Fig. 22. Photoconductivity and absorption spectra of the multilayer structure of polydiace-tylene [150]. Photocurrent value for black points is ten times lower than it is shown in figure

Transport phenomena were intensely studied for various types of the polydiacetylenes [152–171].

Donovan and Wilson [157, 158, 162–164] have come to the conclusion that an extremely high mobility of carriers, greater than 20 $m^2 V^{-1} s^{-1}$, whose velocity saturates at the velocity of sound at field as low as 10^{-2} V m^{-1}, exists. The main arguments were based on the comparison of the amplitude of the pulse photocurrent excited by a 20 ps laser pulse and the magnitude of the charges collected at the electrodes. The anomalies in the assumption underlying their analysis were shown later [166–171]. Regarding the nature of a laser-pulse-induced photocurrent, the recombination of free carriers in a 1D system and the occurrence of the magnetic field effect on the photoconductivity, Frankevich [170, 171] has shown that a normal mobility of about 3×10^{-4} $m^2 V^{-1} s^{-1}$ is typical for polydiacetylenes. The same value was obtained from time of flight experiments [165–169].

The most authors consider that the Onsager theory cannot be applied to the photogeneration of charge carriers in polydiacetylenes and transport processes are controlled by deep traps.

The photoconductivity is highly anisotropic and its value along the macro-molecule chain is 10^3 times larger than in the perpendicular direction. The anisotropy is mainly due to the various values of the mobilities along and transverse to the chains.

Polydiacetylenes have prospective practical applications. Electron beam irradiation is used to make electronic walls intersecting the chains of the single crystal polymer [172]. Photocarriers travel along the polymer chains until they stop at the walls. The carrier motion is consistent with that of the acoustic solitary wave polaron. Photocarriers of opposite sign accumulating at the walls are shown to recombine. It may therefore be possible to use such structures as molecular electronic devices. Photochemical polymerization can be used for preparing holograms. Diffraction efficiency up to 40% on the spatial frequency 1600 mm^{-1} and a sensitivity of 5 $cm^2 y^{-1}$ was realized on polydiacetylene [173, 174].

4.3 Polymers with Triple Bonds

Polymers with triple bonds and heteroatoms in the macromolecule chain were among the first photoconducting polymers [8, 93–96]. The structural formulas for some among nearly seventy polymers investigated may be represented by

$$R_1-(C\equiv C-R-C\equiv C)_n-R_1 \text{ and } R-(C\equiv C)_n-R$$

where R and R_1 are organic radials with functional groups or heteroatoms. Some examples of the polymers investigated are presented in Table 1. The photoresistors made possessed a dark conductivity varying from 10^{-12} to 10^{-14} S cm^{-1} which increased under integral illumination by 2 or 3 orders of magnitude. The

Table 1. Photoconducting polymers with triple bonds [93–96]

N	Formula

1 $C_6H_5-C{\equiv}C-\left(-C{\equiv}C-\bigcirc-C{\equiv}C-\right)_n-C{\equiv}C-C_6H_5$

2 $C_6H_5-C{\equiv}C-\left(-C{\equiv}C-\underset{}{\overset{OH\ \ OH}{[anthracene]}}-C{\equiv}C-\right)_n-C{\equiv}C-C_6H_5$

3 $NO_2-\bigcirc-C{\equiv}C-\left(-C{\equiv}C-\underset{}{\overset{OH\ \ OH}{[anthracene]}}-C{\equiv}C-\right)_n-C{\equiv}C-\bigcirc-NO_2$

4 $[naphthyl]-C{\equiv}C-\left(-C{\equiv}C-[anthracene]-C{\equiv}C-\right)_n-C{\equiv}C-[naphthyl]$

5 $(CH_3)_3-C-C{\equiv}C-\left(-C{\equiv}C-\bigcirc-C{\equiv}C-\right)_n-C{\equiv}C-C-(CH_3)_3$

6 $NO_2-\bigcirc-C{\equiv}C-\left(-C{\equiv}C-\bigcirc-C{\equiv}C-\right)_n-C{\equiv}C-\bigcirc-NO_2$

7 $\left(-C{\equiv}C-\bigcirc-N{=}N-\bigcirc-C{\equiv}C-\right)_n$

8 $\left(-C{\equiv}C-\bigcirc-C{\equiv}C-\right)_n$

$\left(-C{\equiv}C-CH_2O-\bigcirc-\bigcirc-OCH_2-C{\equiv}C-\right)_n$

Table 1. Contd.

N	Formula

9   $CH_2OH—C{\equiv}C—$ $\left(—C{\equiv}C—\underset{CH_3\quad N\quad CH_3}{\bigcirc}—C{\equiv}C—\right)_n$ $—C{\equiv}C—CH_2OH$

10   $(—C{\equiv}C—)_n$

photocurrent dependence on the light intensity can be expressed by the equation $i = \alpha L^n$, where n varies from 0.5 to 1. This may be due to the exponential or continuous distribution of the local levels according to the model proposed by Rose [1].

It was found that preliminary ultraviolet irradiation alters the photoelectrical sensitivity. Figure 23 represents the spectral response curves for the photoconductivity (curve 1) and the photoelectromotive force (curve 3) of poly (p_1p'-diethynylazobenzene).

$$\left(—C{\equiv}C—\bigcirc—N{=}N—\bigcirc—C{\equiv}C—\right)_n$$

The photoconductivity spectrum after preliminary irradiation of polymer is given by curve 2. The observed redistribution of the peaks is partly reversed on prolonged exposure to air. The bathochromic shift of the shorter wavelength peak depends on the exposure time. Ultraviolet irradiation produces a slight change in the polymer colour. Such irradiation increases, likewise, the photoelectromotive force; a 1.5 h irradiation increases it 10 times. The photoconductivity spectrum is situated at longer wavelengths than the photoelectromotive force spectrum.

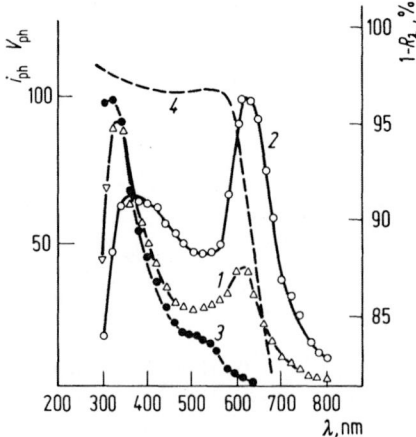

Fig. 23. Spectra: photoconductivity of poly(p,p'-diethynylazobenzene (*1*), the same after 10 min under ultraviolet irradiation (*2*); photoelectromotive force (*3*); absorption in diffuse reflection (*4*) [95]

The presence of a photoconductivity peak at 610 nm at the threshold of the absorption spectrum (curve 4) is a common phenomenon in inorganic semiconductors and is explained by competition between surface and volume recombination processes of the charge carriers. The optical activation energy determined from the spectral photoconductivity threshold is equal to 1.82 ± 0.02 eV. The thresholds of the photoelectromotive force and the absorption spectra are likewise in agreement with this value. It is remarkable that the same value has been found for the activation energy of the dark conductivity in this polymer [175].

The sign of the dominant charge photocarriers in most of the polymers investigated has been found to be positive. The photosensitivity increase under ultraviolet irradiation might be connected with the photoionization of polymer molecules and the creation of local, positively charged centers acting as photoelectron traps. This mechanism seems to be confirmed by the electron spin resonance method.

The photosensitivity and its increase after ultraviolet illumination was confirmed by electrophotographic methods for polymers with triple bonds [176]. Later on, the sensitivity increase under ultraviolet light action was observed for different types of polymers.

By analogy with polyacetylenes, one can consider molecular absorption with Frenkel type exciton formation in the long wavelength region of the spectrum [14]. There is a sharp increase of the absorption and photosensitivity in the short wavelength region. These facts may be connected with the strong optical band to band transitions buried under the excitonic absorption. The definition of the specific gravity of the excitonic and band transition needs detailed research. The photosensitivity change under ultraviolet light has a close analogy with the formation of the color centres in inorganic crystals.

Polymers with triple bonds have been applied in electrophotography [176] and were the first among organic photoconductors with sensitized by dyes photoconductivity [19, 20].

4.4 Poly(phenylacetylenes)

Poly(phenylacetylene) and poly(diphenylacetylene) consist of polyaromatic conjugated fragments side by side with polyene chains. The photosensitivity strongly depends from acceptor concentration, supermolecular structure, the synthesis procedure [177]. The acceptor molecules inclusion increased the photosensitivity, the optimum of which was obtained for heterogeneous phases with a large interface area. The transition of the amorphous structure into the crystal one promoted the photosensitivity increase. The maximum quantum yield of 10^{-3} was at the energy 3 eV, mobilities were varied from 10^{-9} to 10^{-6} $m^2 V^{-1} s^{-1}$.

For poly(phenylacetylene) doped with acceptor iodine molecules the long wavelength bands at 940 nm were ascribed to CT complexes [178, 179]. The band model with trap controlled conductivity was used to the photoconductive properties. Carrier injection into polymers is the main process for dye sensitized photoconductivity.

The detailed analysis of the photoexcited charge transfer states in the poly(phenylacetylene) chloranil structure leading to the photoconductivity was made with the help of magnetic methods [180]. Pulsed photoconductivity with an intrinsic photogeneration threshold in poly(phenylacetylene) was interpreted in the terms of Onsager's theory with quantum efficiency of $2-10^{-2}$ and a thermalization length 2.2 nm [181]. High photosensitivity in irradiated by electron beam cis-poly(phenylacetylene) films [182] and poly [(oxygen trimethylsilylphenil)acetylene] [183] arose mainly from trap-modulated carrier emission and effective transport of holes.

The investigation of the quantum yield of the poly(diphenylacetylene) and its complex with TNF show that maximum photosensitivity was at 25% of the acceptor content [184]. This relates to the case when all the conjugated blocks form complexes with TNF. The strong dependence of the quantum yield on the electric field strength permitted the explanation of the photosensitivity in the framework of Onsager's theory with a thermalization distance of 2 nm.

4.5 Poly(phenylene), Poly(thiophene), Poly(phenylenesulfide)

The chemical structures of the polymers are as follows:

poly(phenylene) $\left(-\langle\bigcirc\rangle-\right)_n$ poly(thiophene) $\left(-\langle S\rangle\langle S\rangle-\right)_n$

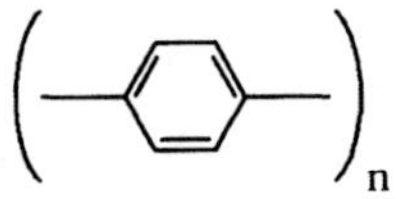

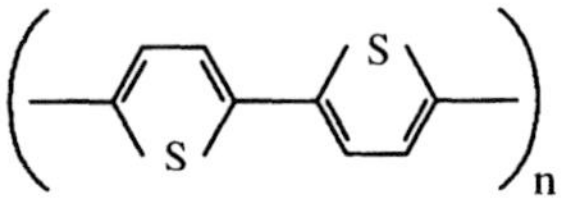

poly(phenylene sulfide) $\left(-\langle\bigcirc\rangle-S-\right)_n$

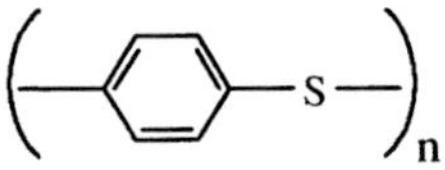

Poly(p-phenylene) is a semiconducting material with well-defined conductivity activation energies of 0.2 eV at temperatures above 240 K and of 0.05 eV at lower temperatures [185]. Conductivity may be approximated by interchain hopping. High quality films may be obtained by the electrochemical method [186]. The conductivity of such films can change from 10^{-12} to $10^2\,S\,cm^{-1}$ upon doping. The ground excited state is situated at 3.4 eV; levels at energies less than 3 eV are related to polaron or bipolaron states. Fast and slow photocurrent components related to singlet and triplet transitions were observed at first [187]. Apparently this can be explained by the corresponding trap limited conditions. The mobility was estimated at $6 \times 10^5\ m^2\,V^{-1}\,s^{-1}$. The activation energy coincides with the intramolecule excitation of the π-electrons. The photoconductivity activation energy was 0.5–1 eV [188]. Polaron formation as in the case of trans-polyacetylene leads to two localized states in the middle of the forbidden gap, which can be interpreted as soliton functions.

Polythiophene is a highly crystalline polymer with the chain analog to cis-polyacetylene. The sulfur atoms stabilize the structure and interacts poorly with

π-electrons of the main chain. The polymer may be considered as a kind of polyene with an undegenerate ground state. So the existence of the stable soliton states becomes impossible. Therefore the polaron states are the most profitable in such structures.

From photoinduced absorption, luminescence and electron spin resonance observations, the dominant photocarriers generated in the polymer were shown to be polarons and bipolarons [189–191]. It was found that the magnitude of photoinduced absorption is rather independent of the condition of sample preparation whereas the photoluminescence intensity is strongly influenced. The results suggest that the luminescent exciton does not play a primary role in the photogeneration of polaronic species.

There are few data concerning the photoconductivity of polythiophene [192,193]. The conductivity change under the combined action of NO_2 and white light is discussed assuming the participation of photo-induced absorption and desorption processes.

Photoelectrochemical cells with polythiophene film as an active electrode and a lead plate as a counter electrode in Pb (ClO$_4$) acetonitrile electrolyte has an open circuit voltage of 0.8 V, short circuit current of 2×10^{-4} A cm^{-2}, efficiency coefficient of 0.03%, fill factor of 15% [194]. The absorption and photosensitivity spectra of such a cell are shown in Fig. 24. The small bathochromic shift in the longwave region for photosensitivity may be related to the photogeneration of the charge carriers via surface states. The photosensitivity maximum is close to the maximum solar intensity. The parameters exceed the ones obtained with polyacetylene.

Heterojunctions of polythiophene with polypyrrole [195] and Cds [196] of the Schottky type were constructed and tested. The height of the barrier was 0.8 eV. The photogeneration of the charge carrier takes place in the depletion layer of the thiophene with consequent separation in the barrier electric field.

Poly(-p-phenylene sulfide) is an electronic material with good prospects due to high thermostability up to 280 °C and has the possibility of changing the resistance up to 19 orders of magnitude with doping.

The absorption edge of the polymer is at 380 nm.

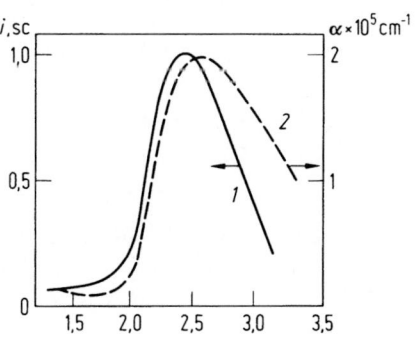

Fig. 24. Short circuit photocurrent (*1*) of the electrochemical cell with polythiophene and its absorption spectra (*2*) [194]

The photoconductivity spectra are situated in the region from 300 to 700 nm have no agreement with the absorption spectra [197–206]. The strong surface recombination in the fundamental absorption band or the presence of impurities may be the reason for this fact. The experimental result is shown in Fig. 25 [200]. The absorption edge is very close to the onset of the photoconductivity. Oxygen treatment leads to an increase in the conductivity and photoconductivity of thousand times. The photo and dark conductivities were reversible for oxygen doping and deoxidation. One can assume a weak CT complex formation with oxygen, which absorbs at wavelengths higher than 400 nm. The treatment of the polymer by other acceptor molecules increases the photoconductivity at the longer wavelength. Some results can be seen in Fig. 26 [202]. The increase of the crystallinity increases the photoconductivity. As for polymers with triple bonds, it was shown the photosensitivity of polyphenylene sulfide increase after preliminary ultraviolent illumination.

The transport of the charge carriers had typical dispersive character and the mobilities of the holes was of the order $10^{-8} \, m^2 \, V^{-1} \, s^{-1}$.

The increase in the mobility [200, 201] and concentration of the charge carriers [202–204] after doping was shown by time of flight methods. The drift mobility for the pure polymer had an activation energy of 0.25 eV. As a rule, the

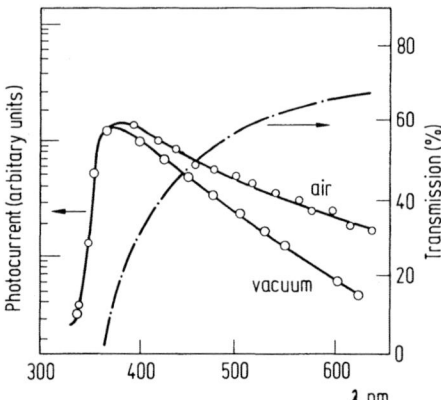

Fig. 25. Absorption and photoconductivity spectra of polyphenylene sulfide in air and vacuum [200]

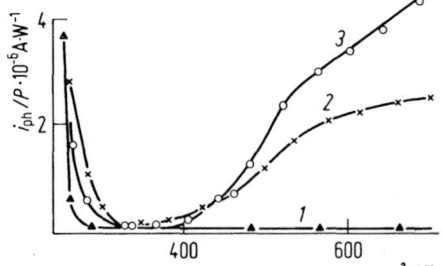

Fig. 26. Photoconductivity spectra of polyphenylene sulfide: initial (1), doped with tetracyanohinodimethene (2) and tetranitrilpiromellit acid (3) [202]

hopping model is preferable for the interpretation of the photoconductivity mechanism of the polymer.

The pristine and doped poly-phenylene sulfide may be used for light energy conversion to the electrical one. The efficiency coefficient is of the order 0.01–0.2% for various lightwaves, fill factor 0.07–0.20. The characteristics are better than ones for doped TNF PVC.

4.6 Polyxylylidenes and Polyxylylenes

The polymers have the next following chemical structure:

Polyxylylidene $\left(-\!\!\left\langle\bigcirc\right\rangle\!-\!CH\!=\!CH-\right)_n$

Polyxylylenes $\left(-CH_2-\!\left\langle\bigcirc\right\rangle\!-CH_2-\right)_n$

The first photosensitivity observation in the polyxylylidenes revealed or underlined the role of chemical structure in photoconductivity [207, 208]. Some results are shown in Fig. 27. For example the bathchromic shifts were observed with CH_3 or CN groups introduction in the macromolecule chain. The presence of a photoconductivity peak at the threshold of the absorption spectrum was explained by competition between surface and volume recombination processes of the charge carriers. A close resemblence between thermal energy activation of the dark conductivity, optical activation energy for photoconductivity, and absorption edge energy was obtained. This means the same electronic transitions take part in these processes. Preliminary ultraviolet illumination increased the photosensitivity due to the formation of the photochemically active centers. Beginning with four monomer links in the macromolecule chain, the absorption, fluorescence and photoconductivity spectrum had the least difference. So the electron transitions of the small conjugated sections are responsible for photophysical processes.

In the last few years a renewal of interest in the poly(p-phenylenevinylene) may be connected with its quasi-one-dimensional structure without degenerated state [209–217]. The molecular structure of this polymer is situated between polyacetylene and polyphenylene. The absorption edge is at 2.7 eV and characterized by direct band transitions [212]. Doping can lead to a conductivity increase of 10^{13} times. The absorption, luminescence, and photoconductivity data may be explained by the participation of polaron and bipolaron states. The according energetic schemes are shown in Fig. 28 for trans-polyacetylene, polyphenylene and polyphenylvinylene. The value of the states with the energy of 0.7 eV from the band edge is very close for all three polymers. This means that in spite of strong difference in the molecule structures nearly identical states are formed due to the deformation of the conjugated chains.

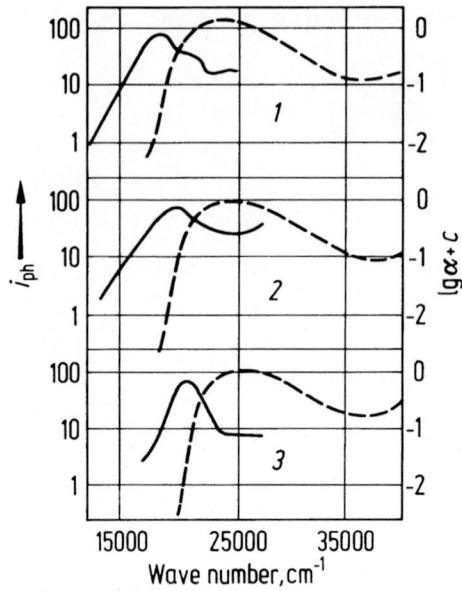

Fig. 27. Diffuse reflection (– – –) and photoconductivity (——) spectra for polymers:

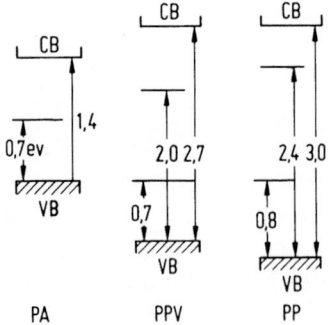

[207]

Fig. 28. Band schemes for *trans*-polyacetylene (PA), poly-phenylvinylene (PPV) and polyphenylene (PP). The energetic intervals are shown in electron Volts [212]

The photoconductivity of polyphenylenevinylene strongly depends on doping and polarization of light for the oriented films. The nature of the large photoconductivity, in the picosecond time interval; observaed by Bradley et al. [210, 211] needs detailed investigation.

Dispersive hole transport in polyarylene-vinylene is caused by hopping between conjugated blocks [215]. The heteroatom inclusion leads to the decrease in the dispersion times. The typical value of the drift mobility is $10^{-9}\,m^2\,V^{-1}\,s^{-1}$ with an activation energy 0.27 eV.

The photoconductivity of poly-p-xylylenes were investigated under various conditions [218–221]. The fundamental edge absorption of the polymers is at 300 nm. The photocurrents in the visual range of the spectra depended on the electrode nature. So it was interpreted as photoinjection of the charges from electrodes and separation of them at a Schottky type barrier. Suppression of hole injection for the plasma-treated polymer is related to the existence of an oxidized surface layer.

4.7 Polyimides, Heteroatom and Heterocyclic Polymers

Polyimides and other heterocycle polymers possess high thermostability and excellent mechanical and electrical properties. That they are also photosensitive makes them highly interesting. Even during the pioneering stages, the photoconductive properties of the polyimide were established [222, 223]. Donor-acceptor interaction between structure elements of the polyimide chains is the main factor determining the photosensitivity. The structural formulas and the main energetic characteristics of some heterocyclic polymers are presented in Table 2.

For most of these polymers the coincidence between the doubled thermal activation energy of the dark conductivity, the energy of the absorption edge and the optical activation energy of the photoconductivity, defined from the photoconductivity spectrum, was established. These facts permit us to make conclusions about the intrinsic nature of the conductivity. The coincidence of the optical activation energy with singlet electronic transition shows that the photogeneration of the charge carriers is caused by singlet transition excitation. The heterocycle position in the chain determines the main electronic properties. Polymers with heterocycles in the main chain demonstrate electronic conductivity, the ones with the heterocycles in the side groups have hole conductivity. The polyconjugation decrease leads to an increase in the forbidden gap intervals and the energy of the excited states. The formation of the inter and intramolecule complexes was proved by various methods.

Photoconductivity of Kapton, (polypyromellitimide where R is oxygen) was investigated in detail [224–234]. Frenkel, Onsager, hopping and other models were used by different authors for explanation of the photoconductive properties. The photoconductivity spectra of Kapton film for various directions of the electric field are presented in Fig. 29 [230]. The high anisotropy depending on molecule orientation and its degree of regularity was established. The main conclusion is that the photocurrent direction between the chains prevails over the direction along the chain. The hopping charge transfer is the main mechanism. The hole mobility $5 \times 10^{-16} \, m^2 \, V^{-1}$ at $50 \, °C$, $10^{-13} \, m^2 \, V^{-1} s^1$ at $200 \, °C$ with thermal activation energies of $0.2 \, eV$ and $0.8 \, eV$ respectively was established [232]. The doping of the Kapton by electron donor molecules increases the photoconductivity by up to five orders of maguitude [234] due to the charge transfer formation. The photogeneration mechanism is in agreement with the Onsager model when the quantum yield is unity, thermalization distance $1.3 \, nm$.

Table 2. Thermal activation energy of conductivity E_d, edge energy of the electronic absorption spectra E_{opt}, and optical activation energy of photoconductivity $E_{\lambda 1/2}$ for some polyheterocycle polymers [222]

R	R'	E_d, eV	E_{opt}, eV	$E_{\lambda 1/2}$, eV
Polypyromellitimides				
$-O-$	$-$	1.30	2.63	2.85
$-S-$	$-$	1.21	2.42	2.40
NH	$-$	1.07	1.98	2.16
$-N(CH_3)-$	$-$	1.06	1.96	
$-N(Ph)-$	$-$	1.03	2.0	
Poly(N-phenyl)benzimidazoles				
		1.47	3.01	$-$
		1.18	2.39	$-$
		1.35	2.74	$-$
		1.50	3.10	$-$
		1.34	2.78	$-$

Table 2. Contd.

R	R'	E_d, eV	E_{opt}, eV	$E_{\lambda 1/2}$, eV
Polybenzoxazoles	$\left[-C\underset{N}{\overset{O}{\diagdown}}R\underset{N}{\overset{O}{\diagup}}C-R'-\right]_n$			
(biphenyl)	(m-phenylene)	1.40	2.82	2.87
	(phenyl-O-phenyl)	1.42	2.86	2.97
	(phenyl-O-phenyl-O-phenyl)	1.45	2.86	2.96
	(phenyl-S-phenyl-S-phenyl)	1.42	2.86	2.95
(dimethylphenyl-O-dimethylphenyl)	(m-phenylene)	1.47	3.17	3.10

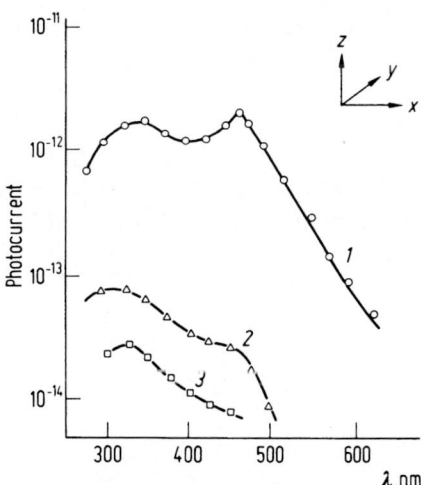

Fig. 29. Photoconductivity spectra of Kapton films at an electric field strength of 5×10^5 V·cm^{-1} for various directions indicated. _1_ – along z axis; _2_ – along y axis; _3_ – along x axis. The z axis is perpendicular to the film plane in which the molecules are oriented [230]

Compared to Kapton, more sensitive soluble polyimide films were investigated [235-236]. A photosensitivity of up to 30 m^2 J^{-1} was achieved in electrophotographic regimes. The photogeneration of the charge carriers occured via exiplex formation and the holes are the predominant carriers. The spectral dependence on the quantum yield is shown in Fig. 30. The close resemblance of the experimental data for quantum yield and its dependence on the electrical field with theoretical curves for hopping generation model was established. The decay kinetics in intervals less than 10^{-2}s are typical for dispersive transport. Longer times were related to the deep trapping in the polymer structure. The irregularity of the molecule centers in the chain is one of the reasons for dispersive transport and low mobility.

New types of polymers containing imide groups were proposed recently with high photosensitivity [237, 238] of the order less than 1μJ. Photosensitivity of the polymers composed of a series of thiophenylenes moieties and imide groups [238] remarkably increased with the increase of crystallinity. The overlapped π-orbitals perpendicular to the polymer chains lead to the more extended conjugated system.

Individual investigations of different types of heteroatom and heterocycling polymers have been carried out [14]. Among them, photosensitivity was observed in polyanilines [239–241], diphenylaminsulfide [242], poly[N-benzyldiphenylamino methane] [243], polyorganophosphazenes [244], polymeric dyes [245–247], polyquinoxaline [248], poly(paracyclopropane) [249] and others [250–253].

From the various types of heteroatom- and heterocycle-containing photoconducting polymers apparently only polyimides can be considered as polymers for which there is information about applications. It was proposed photoconducting polyimide be used as a photosensitive element in liquid crystal spatial light valves [254]. The simplest modulator structure consists of thin films of photoconductor and liquid crystal sandwiched between the transparent conductive electrodes. Controlled voltages are applied to the electrodes. Inorganic photoconductors were used in such optically addressed modulators until recently. High mobilities of the charge carriers in such photoconductors hinder the high spatial resolution needed for such devices. As we already know, very low mobilities are inherent to organic polymers. So it allows us to obtain high spatial resolution. Besides high mechanical and electrical stability, the possibility of chemical and spectral sensitization of the photoconductivity opens real perspectives for spatial light modulators with polymer photoconductors in optical

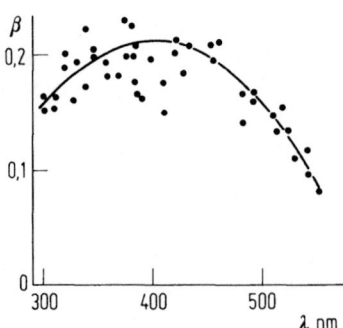

Fig. 30. Quantum yield of the charge carrier formation versus the light wavelength for polyimide [235]

information processing, holography, optical interconnection, and laser light modulation.

Sensitized for blue-green or red light, photoconductive polyimides and liquid crystal mixtures of cyanobiphenyls and azoxybenzene have been used in spatial light modulators [255–261]. Modulation procedure was achieved by means of the electrically controlled birefringence, optical activity, cholesteric-nematic phase transition, dynamic scattering and light scattering in polymer-dispersed liquid crystals.

Photosensitivity spectra of the modulator are shown in Fig. 31 [259]. The positive sign of the dominant charge carriers in polymer seems to be confirmed by comparing the photocurrent values at different bias. The existence of the rectifying barrier at the photoconductor – liquid crystal interface follows from the shift of the main maximum of about 30 nm with the polarity change. The existence of such a barrier was confirmed by current voltage characteristics which were typical for rectifying structures and a photoelectromotive force of 0.3 eV the sign of which was the same no matter which modulator side was lit. The volt-contrast and current voltage characteristics for a modulator with cholesteric nematic phase transition is represented in Fig. 32 [254]. The structure scatters the light in the initial stage thanks to the confocal texture of liquid crystal. The spiral unwinding at high voltages leads to the transparent state. Both types of characteristics at the initial stage are unsymmetrical versus polarity bias due to the interface barrier.

The operating regimes strongly define the modulator characteristics. The pulse record regime allows us to obtain the highest parameters for response time and resolution. Diffraction efficiency and response times versus spatial frequency for commutational and pulse regimes are shown in Fig. 33 and Fig. 34 accordingly. The best results obtained for the pulse writing regime are the following:

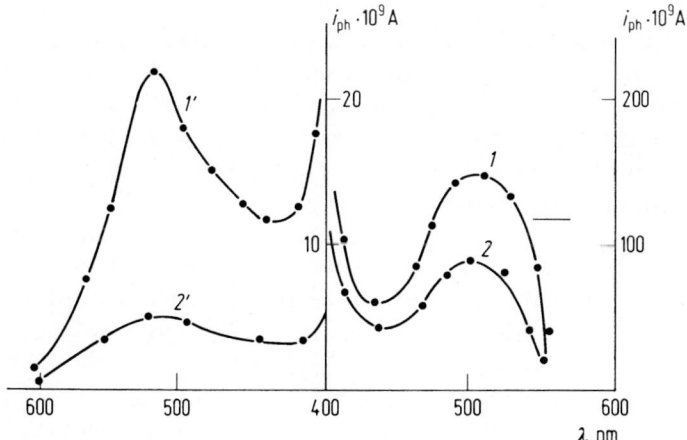

Fig. 31. Photosensitivity spectra of the photoconductive polyimide-liquid crystal modulator for positive (*right*) and negative (*left*) bias on the photoconductor. Voltages in V 80 (*1,1'*); 20 (*2,2'*) [259]

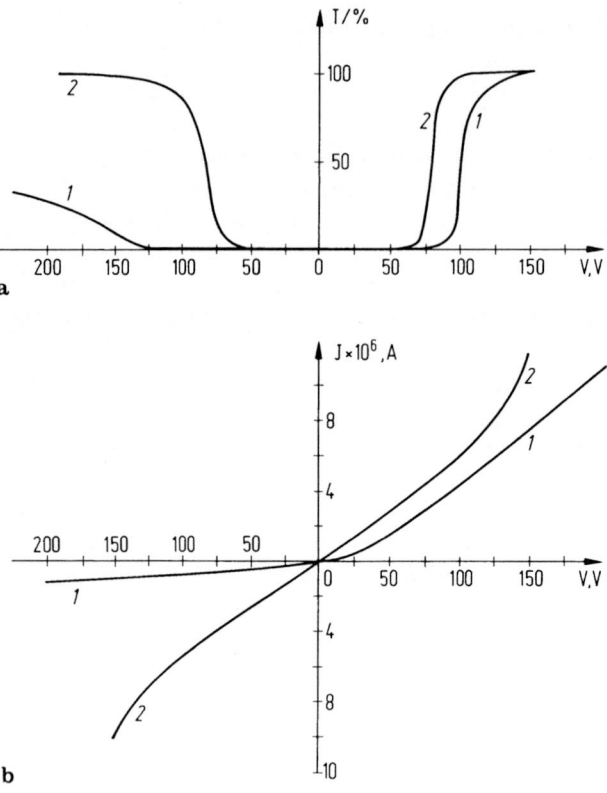

Fig. 32a, b. Volt-contrast (**a**) and current-voltage (**b**) characteristics for polyimide modulator with cholesteric-nematic phase transition without (*1*) and with switch (*2*) on the recording light [254]

limiting resolution up to 1500 mm^{-1}, sensitivity 10^{-8} J/cm^2%, response time $\geq$ 100 ms, maximum diffraction efficiency 36%. These parameters are close to the theoretical ones for thin flat holograms and the modulator can be considered as a hologram. Such high resolution is not reached for modalators with inorganic photoconductors. The parameters obtained may be lightly changed by variation of the thin films thicknesses and the value of the dielectric permittivity of the liquid crystal. The results are presented in Fig. 35. It may be remarked that the use of the polymer photoconductor itself for the alignment of the liquid crystal simplifies the modulator production. The manufacture of the device with high information capacity becomes possible due to the ease of increasing the aperture compared with using inorganic photoconductors. The modulators with more than 100 mm aperture have been manufactured and tested [260].

The modulators mentioned above contained liquid crystal molecules. An entirely solid state device is naturally preferable in practice. The polymer-dispersed liquid crystal materials may be prospective counterparts of the modulator. One of the modes of operation of such thin films are as follows. The

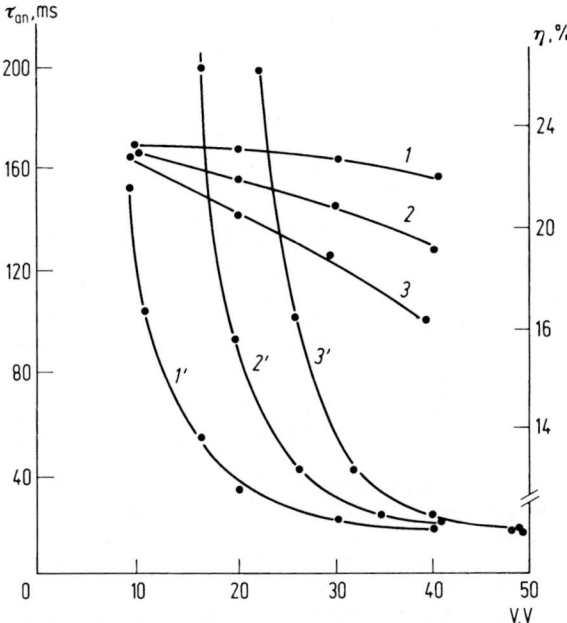

Fig. 33. Diffraction efficiency versus voltages for spatial frequencies in mm^{-1} 10 (*1*), 28 (*2*), 51 (*3*). Switch on time versus voltages at spatial frequencies in mm^{-1} 10 (*1'*), 36 (*2'*), 60 (*3'*). Commutational regime for polyimide-liquid crystal modulator with controlled birefringence [256]

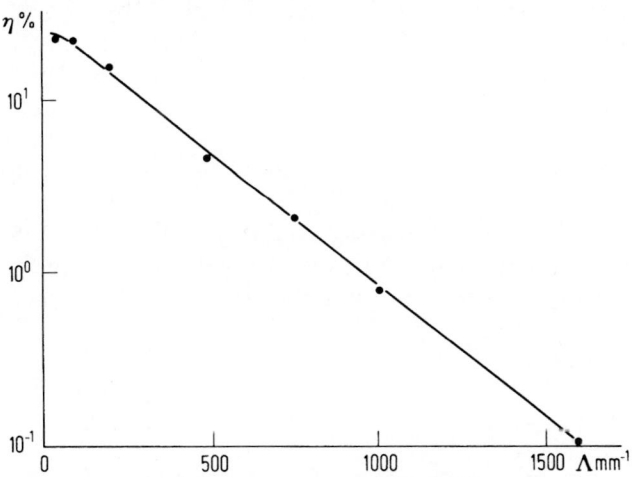

Fig. 34. Diffraction efficiency versus spatial frequency for electrically controlled birefrigence liquid crystal-polyimide modulator in pulse regime [257]

nematic droplets with positive dielectric anisotropy are randomly oriented in the polymer binder. In this opaque state the film has a translucent white appearance because of the light scattering. On applying the electric field the droplets align in a direction parallel to the field. If the refractive index matches the refractive index

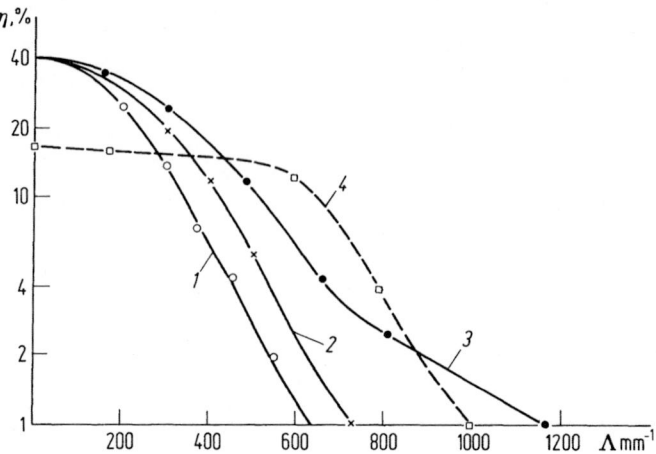

Fig. 35. Modulation function of the polyimide modulator with thickness of the liquid crystal in µm 20 (*1*), 15 (*2*), 10 (*3*), 3 (*4*). Anisotropy of the dielectric permittivity of 12 (*1, 2, 3*); 0.05 (*4*). Polyimide thickness – 2 µm. Pulse regime [14]

of the polymer matrix, the film will pass into the transparent state. Upon the removal of the field the film returns to its opaque state. Submillisecond switching times can be obtained with these films.

The polymer-dispersed liquid crystal was used as a modulating medium in optically controlled modulators instead of the liquid crystal [261–264]. The sandwiched structure from polyimide photosensitive film and the polymer dispersed liquid crystal film – i.e. the optically controlled solid state modulator – had the characteristics presented in Fig. 36 [261]. Contrast ratio 35:1, response time ≥ 400 µs, decay time 80 ms and sensitivity 5×10^{-5} J cm^{-2} were obtained.

Besides polyimides, photoconductive polymers with conjugated bonds also can be successfully used in liquid crystal spatial light modulators [262–264].

The high resolution, mechanical and electrical stability and possibility for sensitization of the photosensitivity make the use of organic polymer photo-conductors in spatial light modulators very attractive. Devices such as phase

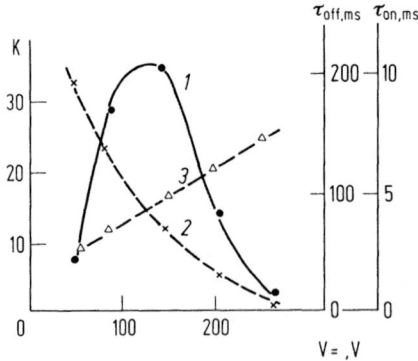

Fig. 36. Contrast ratio (*1*), switch on (*2*) and switch off (*3*) times versus voltages for modulator with polymer dispersed liquid crystal. Photoconductor-polyimide film. Pulse regime [261]

reversible materials, can be used for input and output of the optical information, in optoelectronics, holography and so on.

5 Organometallic Polymers

Various types of bond in organometallic compounds permit us to assume the existence of very interesting semiconducting properties in such materials. First, photosemiconductive properties in organometallic polymers were observed in the early 1960s [93, 265]. Now almost eighty compounds are known in which the photosensitivity has been observed [10, 14]. These are polymers with double and triple bonds containing atoms of silicon, germanium, tin, lead in the macromolecule chain and organoacetylenides of silver and copper. The detailed investigation of the photosensitive properties was made for organoacetylenides, the photoelectrical properties of which were first established in 1963 [265]. Nearly forty ethynyl-metal compounds of the general structure $(R-C\equiv C-M)_n$ or $(M-C\equiv C-R-C\equiv C-M)_n$, where R is an organic radical with or without functional groups and M is a metal (Ag, Cu) were investigated (Table 3) [266–282]. All compounds may be regarded as complicated coordination-type polymers consisting of zig-zag chains of the metal atoms coordinated with radical substituents [266–270]. The structure of the most photosensitive compound-copper phenylacetylenide (PAC) is presented in Fig. 37. The stretching frequency of the disubstituted triple bonds in the infrared spectra equal 2190–2960 cm^{-1} is significantly reduced at 250 cm^{-1} with small dependence on the substituent nature.

Table 3. Structural formulas of the monomeric links of copperorganoacetylenides [14]

N	Formula
1	$CuC\equiv C-(CH_2)_3-CH_3$
2	$CuC\equiv C-CH_2OH$
3	$CuC\equiv C-CH=CH_2$
4	$CuC\equiv C-\underset{\underset{CH_3}{\mid}}{C}=CH_2$
5	$CuC\equiv C-\hspace{-0.3em}\bighexagon$
6	$CuC\equiv C-\hspace{-0.3em}\bighexagon\hspace{-0.3em}-CH_3$

Table 3. Contd.

N	Formula

7    $CuC{\equiv}C$—⟨benzene⟩—Cl

8    $CuC{\equiv}C$—⟨benzene⟩—I

9    $CuC{\equiv}C$—⟨benzene⟩—$COCH_3$

10   $CuC{\equiv}C$—⟨benzene⟩—CHO

11   $CuC{\equiv}C$—⟨benzene⟩—NO_2

12   $CuC{\equiv}C$—⟨benzene⟩—$C{\equiv}C$—⟨benzene⟩

13   $CuC{\equiv}C$—$C{\equiv}C$—⟨benzene⟩

14   $CuC{\equiv}C$—CH_2—⟨benzene⟩

15   $CuC{\equiv}C$—$\underset{\overset{\|}{O}}{C}$—⟨benzene⟩

16   $CuC{\equiv}C$—⟨benzene⟩—⟨benzene⟩

17   $CuC{\equiv}C$—⟨naphthalene⟩

18   $CuC{\equiv}C$—⟨naphthalene⟩

19   $CuC{\equiv}C$—⟨anthracene⟩

Table 3. Contd.

N	Formula

20    $CuC\equiv C-CH_2-O-$⟨benzene⟩$-C-$⟨benzene⟩ ... $O=O$

21    $CuC\equiv C-$⟨pyridine, N⟩

22    $CuC\equiv C-$⟨pyridine, N⟩

23    $CuC\equiv C-N$⟨carbazole⟩

24    $CuC\equiv C$⟨pyrene⟩

25    $CuC\equiv C-C\overset{CH_3}{\underset{CH_3}{-}}CH_3$

26    $CuC\equiv C$⟨ferrocene, Fe⟩

27    $CuC\equiv C-CH$ \\O/ $B_{10}H_{10}$

28    $CuC\equiv C-$⟨benzene⟩$-C\equiv C-Cu$

29    $CuC\equiv C-$⟨benzene⟩ ⟨benzene⟩$-C\equiv C-Cu$

30    $CuC\equiv C-$⟨benzene⟩$-O-$⟨benzene⟩$-C\equiv C-Cu$

31    $CuC\equiv C-CH_2-O-CH-O-CH_2-C\equiv C-Cu$
      $\underset{CH_3}{|}$

Table 3. Contd.

N	Formula

32    $CuC{\equiv}C-CH_2-O-C_6H_4-\underset{\underset{CH_3}{|}}{\overset{\overset{CH_3}{|}}{C}}-C_6H_4-O-CH_2-C{\equiv}C-Cu$

33    $CuC{\equiv}C-CH{=}CH_2$ + $CuC{\equiv}C-$⟨C₆H₅⟩

34    $CuC{\equiv}C$ + $CuC{\equiv}C-$⟨C₆H₅⟩

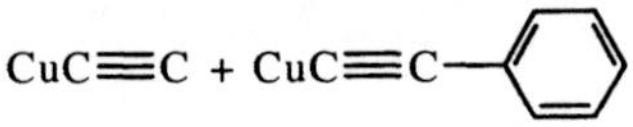

35    $(PhC{\equiv}CH + PhSH) + Cu$

36    $(PhN{\equiv}N-NHPh + PhC{\equiv}CH) + Cu$

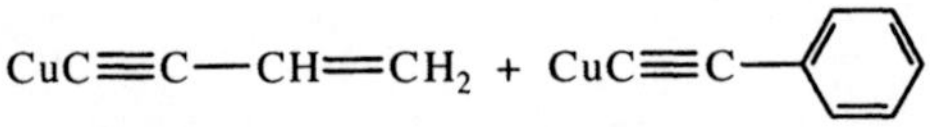

37    $\begin{array}{ccc} Ph-N & \cdots\,Cu- & N-Ph \\ \| & & \| \\ N & & N \\ \backslash & & / \\ Ph-N-Cu & \cdots & N-Ph \end{array}$

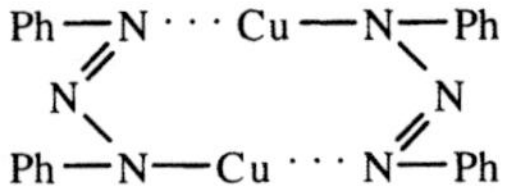

38    $Ph_3PCuC{\equiv}CPh$

39    $(C_2H_5)_2PhPCuC{\equiv}CPh$

40    $PhC{\equiv}CH \cdot CuCl$

41    $CuC{\equiv}C-O-$⟨C₆H₅⟩

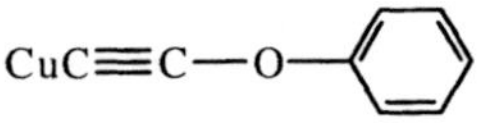

42    $CuC{\equiv}C-S-C_2H_5$

43    $CuC{\equiv}C-S-$⟨C₆H₅⟩

44    $CuC{\equiv}C-P\Big\langle{\overset{C_4H_9}{C_4H_9}}$

45    $CuC{\equiv}C-P\Big\langle$ (two C₆H₅ groups)

46    $CuC{\equiv}C-As\Big\langle$ (two C₆H₅ groups)

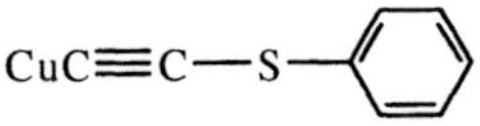

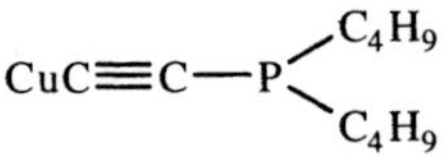

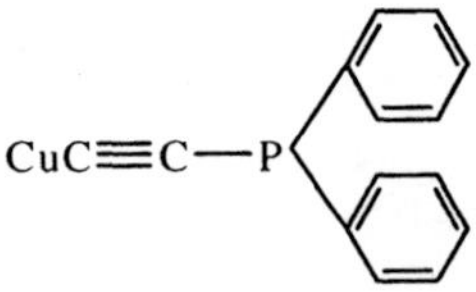

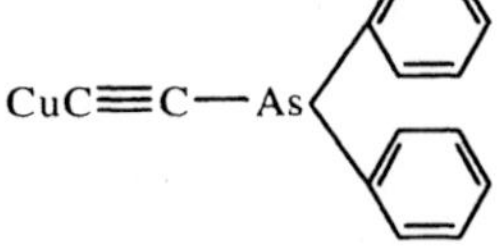

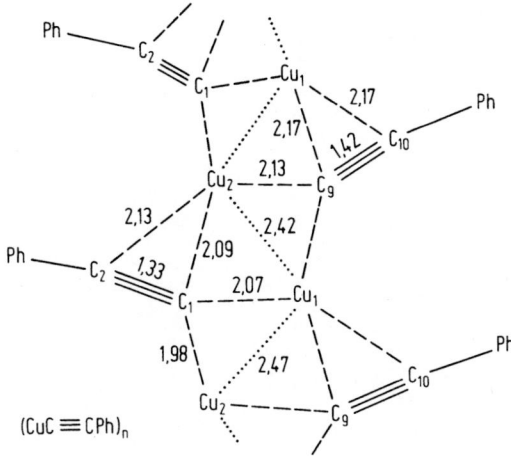

Fig. 37. Structure of polycopperphen-
ylacetylenides (CuC≡CPh)$_n$. The
length of the links is in Å [14]

This means the coordination of the copper *d*-orbital with the loss of the orbitals of at least two acetylenic groups. Photoelectron X-ray spectroscopy confirms this structure [268]. The same results were confirmed by the external photoemission method [269]. The photoelectrical work functions for copper acetylenides with different substituents are in the range 5–5.4 eV. This means that the upper filled state is conditioned by the CuC≡C structure motif.

5.1 Optical and Photoelectrical Properties

The sign of the dominant photocarriers is positive for all acetylenide polymers. The photoelectrical work function, absorption edge energy, optical activation energy of the photosensitivity and electron affinity for some polyacetylenides of copper are shown in Tables 4 and 5. Optical absorption and photoconductivity spectra are presented in Fig. 38 [270–274]. One can see that both spectra reveal some maxima, the origin of which is evidently the same and explained by an intrinsic electronic transition in these polymers. The photoconductivity spectra are shifted to the longer wavelength by about 40 nm compared with absorption ones. The main photoconductivity peaks are situated at the absorption edge – this apparently relates to the strong surface recombination in the region of the fundamental absorption. Little photosensitivity can be observed up to the near infrared region. The nature of such photosensitivity may be connected with structure disorder leading to the broad spectrum of the traps in the forbidden gap.

The common features in the absorption and photoconductivity spectra may be summarized as follows. The increase of the conjugation along the monomeric link leads to the bathochromic shifts to the longer wavelengths. The insertion of the aromatic rings is characterised by the appearance of the vibration structures, proved by the calculation analysis. The spectra may be explained from the

Table 4. Edge energy of the electron absorption spectra ΔE and optical activation energy of photoconductivity $\Delta E_{\lambda 1/2}$ of copper organoacetylenides [14]

N	Formula	Absorption		Photoconductivity	
		λ, nm	ΔE, eV	λ, nm	$\Delta E_{\lambda 1/2}$, eV
1	$CuC\equiv C-C_4H_9$	420	2.73	440	2.72
2	$CuC\equiv C-CH=CH_2$	435	2.5	465	2.41
3	$CuC\equiv C-CH_2-\bigcirc$	415	2.69	445	2.65
4	$CuC\equiv C-C\equiv C-\bigcirc$	465	2.24	500	2.22
5	$CuC\equiv C-\bigcirc$	465	2.45	500	2.38
6	$CuC\equiv C-\bigcirc-C\equiv C-\bigcirc$	475	2.3	510	2.26
7	$CuC\equiv C-\bigcirc-\bigcirc$	468	2.37	–	–
8	$CuC\equiv C-$ (naphthalenyl)	467	2.30	507	2.22
9	$CuC\equiv C-$ (naphthalenyl)	470	2.30	512	2.20
10	$CuC\equiv C-$ (anthracenyl)	530	2.14	550	2.14
11	$CuC\equiv C-O-\bigcirc$	485	2.13	520	2.12
12	$CuC\equiv C-S-C_2H_5$	485	2.32	530	2.2
13	$CuC\equiv C-S-\bigcirc$	450	2.42	530	2.08
14	$CuC\equiv C-\bigcirc-C\equiv C-Cu$	512	2.10	530	2.13

Table 4. Contd.

N	Formula	Absorption		Photoconductivity	
		λ, nm	ΔE, eV	λ, nm	$\Delta E_{\lambda\,1/2}$, eV
15	CuC≡C—⟨◯⟩—⟨◯⟩—C≡CCu	470	2.00	540	2.14
16	CuC≡C—⟨◯⟩—O—⟨◯⟩—C≡CCu	550	2.00	505	2.30
17	CuC≡C—CH=CH$_2$ + CuC≡C—⟨◯⟩	–	–	520	2.24
18	CuC≡C + CuC≡C—⟨◯⟩	540	1.88	620	1.92
19	(PhC≡CH + PhSH) + Cu	–	–	465	2.42
20	CuC≡C—P(C$_4$H$_9$)(C$_4$H$_9$)	360	2.75	440	2.77
21	CuC≡C—P(⟨◯⟩)(⟨◯⟩)	370	2.74	450	2.78
22	CuC≡C—As(⟨◯⟩)(⟨◯⟩)	570	1.88	620	1.75

position of the donor-acceptor resonance interaction between *d*-electrons of the copper atoms and π-electrons of the ligands. Really the compounds of the type HC≡CR and single valent copper ones absorb in the ultraviolet range of the spectrum. The bathochromic shift with the respective substituents in the ethynyl ligand group and high magnitude of the absorption coefficient ($\sim 10^4$cm^{-1}) also confirms the charge transfer between *d* and π orbitals of copper atom and ligands accordingly.

From the point of view of the conjugation system and color theories the copper atoms may be regarded as substituents in the main chromophore with realization of *d*–π conjugation.

Table 5. Photoelectrical work function φ, edge energy of the electron absorption spectra ΔE, optical activation energy of photoconductivity $\Delta E_{\lambda\,1/2}$, and electron affinity, defined from absorption χ_{ab} and photoconductivity χ_{ph} data for copper organoacetylenides [14]

N	Formula	φ, eV	ΔE, eV	$\Delta E_{\lambda\,1/2}$, eV	χ_{ab}, eV	χ_{ph}, eV
1	$CuC{\equiv}C{-}C_4H_9$	5.25	2.73	2.72	2.52	2.53
2	$CuC{\equiv}C{-}$⟨phenyl⟩	5.24	2.40	2.38	2.84	2.86
3	$CuC{\equiv}C{-}C{\equiv}C{-}$⟨phenyl⟩	5.41	2.24	2.22	3.17	3.19
4	$CuC{\equiv}C{-}$⟨phenyl⟩${-}C{\equiv}C{-}$⟨phenyl⟩	5.24	2.30	2.26	2.94	2.98
5	$CuC{\equiv}C{-}$⟨naphthyl⟩	5.22	2.30	2.22	2.92	3.00
6	$CuC{\equiv}C{-}$⟨anthracenyl⟩	5.03	2.14	2.14	2.89	2.89
7	$CuC{\equiv}C{-}$⟨phenyl⟩${-}C{\equiv}C{-}Cu$	5.35	2.10	2.13	3.25	3.22
8	$CuC{\equiv}C{-}$⟨phenyl⟩${-}O{-}$⟨phenyl⟩${-}C{\equiv}CCu$	5.37	2.00	2.30	3.37	3.07

An analysis of the spectra from the solid state physics position has been made [267]. It was shown the energies of the direct transition, obeyed the formula $K = A(h\omega - \Delta E_d)^{1/2}$, and are in the range from 2.46 eV to 2.75 eV. The experimental results are shown in Fig. 39.

The thin control of the energetic characteristics and the electronic transition nature may be carried out by chemical structure alteration. For example the appearance of CH_2 group between phenyl ring and triple bond (compounds 3 and 5 in Table 4) interrupts the interaction between π-electrons of the both elements. This leads to the short wavelength shift in the absorption and photoconductivity spectra. The insertion of the ethynyl group instead of a CH_2 group restores the conjugation. The increase in the numbers and sizes of the

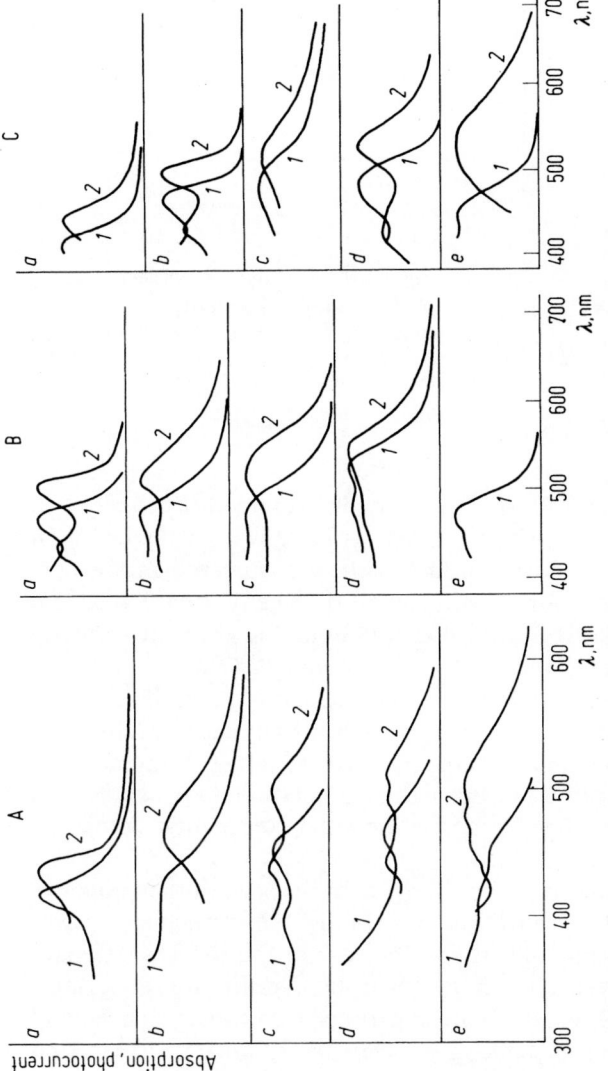

Fig. 38.A–C. Absorption (*1*) and photoconductivity (*2*) spectra of copper organoacetylenides (CuC≡C–R)ₙ [14]
A R is: *a* – C₄H₉; *b* – CH=CH₂; *c* – Ph; *d* – C≡CPh. **B** R is:

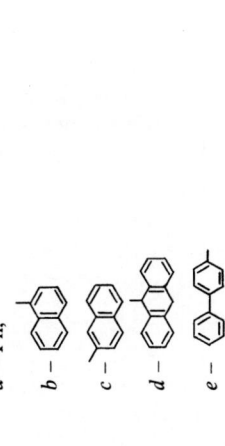

a – Ph;

b –

c –

d –

e –

C R is: *a* – CH₂Ph; *b* – Ph; *c* – OPh; *d* – SC₂H₅; *e* – SPh

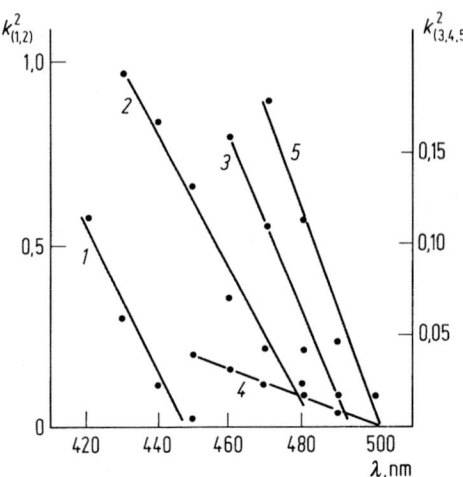

Fig. 39. Absorption coefficient versus photon energy for direct transitions in poly-copperorganoacetylenide $(RC \equiv CCu)_n$ [267] R is: $1 - n\ C_4H_9$; $2 - CH = CH_2$; $3 - Ph$; $4 - C \equiv CPh$; $5 - 1,4\text{-}C_6H_4 \equiv CPh$

aromatic rings in the ligand (N 5, 7–10, Table 4) leads to the bathochromic shift. Strong nonadditive interaction of the aromatic π-electrons with the common $CuC \equiv C$ motif was established. Bathochromic shift was observed in the next order $Cl < I < COCH_3 < NO_2$ for substituents in the phenyl ring. Sulfur and oxygen atoms increase the electron delocalization along the chain. It is clearly seen comparing the spectra of the compounds 5, 11 and 12 (Table 4).

The nature of the substituents in the compounds containing metalloids in the ligand (N 20–22, Table 4) influences the absorption and photoconductivity spectra. The phosphorus atom interrupts the conjugation along the monomeric link. In contrast, the arsenic atom promotes the conjugation. One can assume the more definite dissociation of π-d bonds in the latter compounds compared with the usual acetylenides.

It was remarked earlier that in the visual range of the spectra the bands are due to the charge transfer between d and π orbitals. Simultaneously it was established that there is a strong absorption increase in the short wavelength region. The direct transition existence in the absorption spectra was confirmed: so apparently the charge-transfer bands are superimposed on the valence-band to conduction-band transition in the long wavelength region. The absorption and photoconductivity are defined by the competition between such types of transitions.

Besides photoconductivity, photo-electro-motive (e.m.f.) forces were observed in copper polyacetylenides. The common results characterize the hypsochromic shift of the photo-emf spectra compared to the photoconductivity ones. The maximum of the photoconductivity spectra coincide with the minima of the photo-emf spectra. The optical activation energies of the photosensitivity defined from the photoconductivity and photo-emf spectra are in close agreement and equal to the energy of the absorption edge.

The molecular character of the absorption, photoconductivity and luminescence spectra points out that Frenkel type excition formation takes place at the first stage of the photogeneration of the charge carriers. The dissociation of such excitons on defects or another excitons leads to the free charge carrier formation.

The powerful role of the exitonic migration was proved on the basis of the luminescence and photosensitivity investigations [270]. The preliminary ultraviolet illumination of PAC increases the photosensitivity and decreases the luminescence. The experimental data are given in Fig. 40. One can see the redistribution of the maxima intensity in the spectra without changing their positions. Apparently ultra violet illumination promotes the photolysis of the weak coordination bonds. This leads to the changing of the polymer homolog content. Stimulated exciton dissociation on the ruptured bonds results in an increase in the photosensitivity and a luminescence decrease. The experiments carried out at 77 K show that in the luminescence spectrum of irradiated frozen PAC a new maximum appears with a position close to the phosphorescence maximum of diphenylbutadiene. So the rupture of weak coordination bonds under ultraviolet irradiation was proved.

Simultaneously, it was shown that surface condition strongly influences the photoelectrical properties [275, 276]. Evacuation of the air leads to an increase in the dark conductivity and the photoconductivity (Fig. 41). Dry O_2 reversibly depresses the dark conductivity and the photoconductivity and, equally, the photo-emf without producing a significant change in the spectral distribution. This disproves any photochemical reaction of polymer with oxygen. The activation of oxygen by a discharge enhances the effect described. Oxygen as an electron acceptor should normally increase the photoconductivity of a p-type semiconductor. The tentative explanation of its abnormal behavior

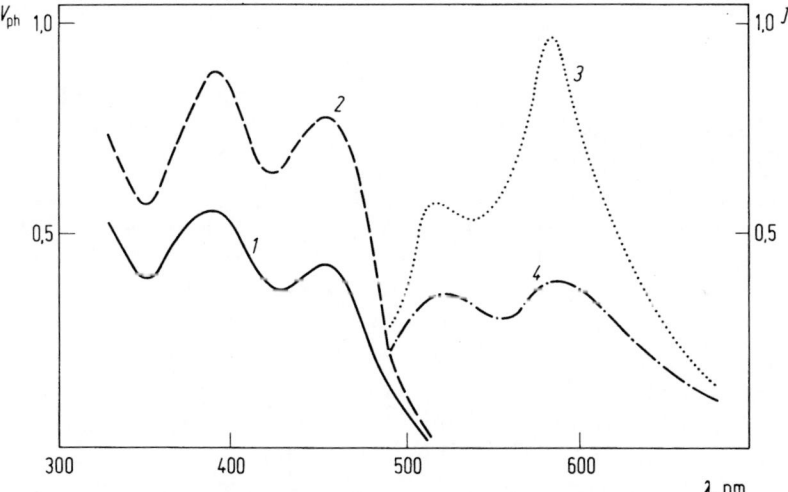

Fig. 40. Photoelectromotive force (*1,2*) and luminescence (*3,4*) spectra of polycopperphenylacetylenide before (*1,3*) and after (*2,4*) preliminary ultraviolet illumination [270]

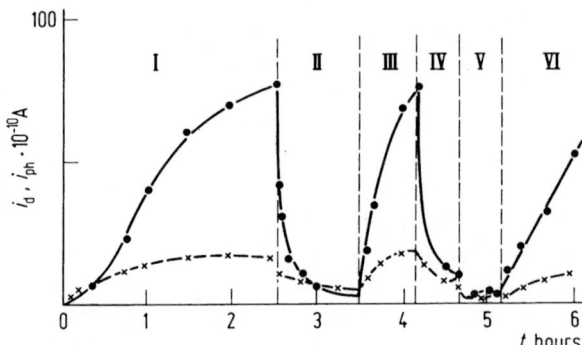

Fig. 41. Oxygen influence on the conductivity (——) and photoconductivity (– – –) of the poly-copperphenylacetylenide (275) *I* – evacuation of the air, *II* – admission, **III** – evacuation, **IV** – oxygen admission, *V* – discharge in oxygen, *VI* – evacuation

here is that adsorbed oxygen creates new recombination surface centers, decreasing the lifetime of the majority of the carriers, i.e. holes.

By means of special experiments in vacuum, reversible photodesoption of oxygen under ultraviolet irradiation was shown [275]. Therefore the electric double layer on the polymer surface is modified. The increase of the emf obtained and the decrease in the dark conductivity and the photoconductivity with the action of water vapor confirms the proposed model [276].

The adsorption of electron acceptors (quinone, chloranil) from the gas phase does not substantially influence the photo-emf of PAC but decreases the dark conductivity and the photoconductivity. The same compounds, however, adsorbed on certain polyacetylenides from solution, increase the photo emf without causing any appreciable change in the spectral distribution. Mercury vapor depresses reversibly the dark conductivity and photoconductivity [276–278].

The results described demonstrate the importance of the surface conditions. The surface states certainly play an important role in the sensitization of the photoeffect in organic polymers.

Appealing to the charge transfer mechanism, in polymetal acetylenides, the mobilities of holes equal to $10^{-5}\,m^2\,V^{-1}\,s^{-1}$ and electrons 1.8×10^{-5} $m^2\,V^{-1}\,s^{-1}$ were obtained for PAC by the time of flight method [14]. The linear dependencies of the transit time versus reverse voltage (Fig. 42) and those for pulse amplitude versus electric field strength show the independence of the mobilities from the electric field. The low mobilities indicate the narrow bands in the organo-metallic polymers and one can consider the hopping mechanism of the charge transfer. The thermal activation energy for mobility was found to be of the order 0.2 eV and trap levels in the interval 0.07–0.1 eV united in the narrow band, the width of which was 3×10^{-1} eV [14]. These facts permit us to conclude that charge transfer in organometellic polymers consists of a series of thermally activated hopping with the participation of the trapping levels. The

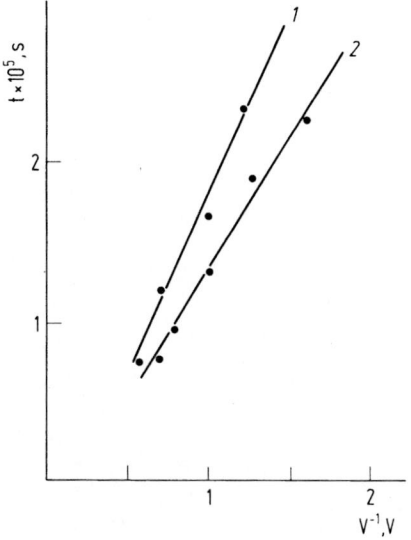

Fig. 42. Dependence of the transit time for holes (*1*) and electrons (*2*) on the voltage for polycopperphenylacetylenide [14]

existence of such levels was also confirmed by the photoconductivity investigation under the action of strong laser light [277].

The transport of charge carriers is evidently also facilitated by the lowering of the barrier potentials between molecules due to the intermolecular bridges in π complexes of copper with acetylenic bonds.

Spectral research permits us to obtain additional information about the role of the molecular structure in photoelectrical properties. The photosensitivity of the copper acetylenides is higher than in polymers with triple bonds without copper atoms. So the coordinating metal atoms and π-electrons of the acetylenic bonds play an important role in increasing the photosensitivity.

Simultaneously, in addition to the main chain, the decisive factors for charge transfer seem to be sterical factors, conditioned by the size of the ligands. So the photosensitivity is diminished in the series with phenyl, naphtyle, and antracenyle substituents (compounds 5, 17, 18,19, 23, 24; Table 3). The same dependencies were established for compounds with heterocycle substituents in the ligand. So, in spite of increasing the number of π-electrons and their delocalization degree, the photosensitivity decreases. The effectivness of the light absorption changes is very small. One can assume the main reason for the decrease in photosensitivity is the difficulties of charge transfer due to the large volume of the substituents.

Aromatic substituents are flat and one can expect compact packing facilitating the charge transfer. Therefore in compounds with substituents which are not flat photosensitivity has to be decreased. It was actually proved for polymers with bulky side groups in the ligand. Moreover, the photosensitivity was not observed in compounds 25, 26, 27 (Table 3) with bulky side groups. The same results were obtained for complexes 38 and 39 and mixed acetylenides 36, 37

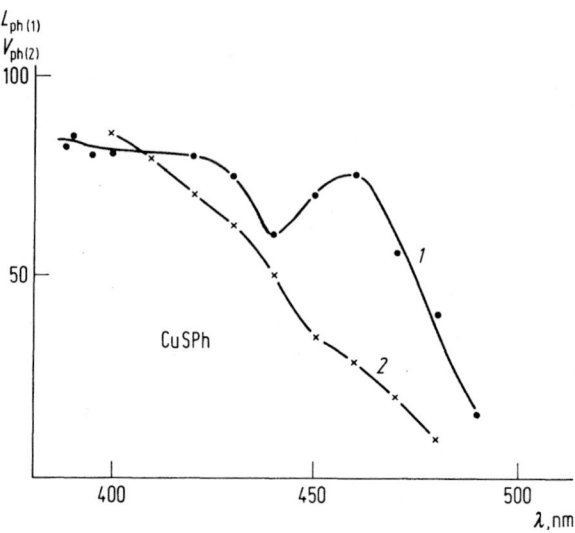

Fig. 43. Photoconductivity (*1*) and photoelectromotive force (*2*) spectra of (CuSPh)ₙ [14]

where the flat structure does not exist. So the inter and intramolecular interactions (short order in the terminolgy of inorganic semiconductor physics) define the charge transfer in metal acetylenides.

In favour of the above mentioned arguments concerning the role of the steric factors one can bring in the results obtained for compounds of the type CuSR, where R-butyl, phenyl, etc. [14]. Physical and chemical properties of these compounds are very similar to the metal organic acetylenides. So the same can be expected for the photosensitive properties. The actual photoelectrical properties were demonstrated in CuSR [14]. The photoconductive and photo-emf spectra of CuSPh are shown in Fig. 43. The optical activation energy is equal to 2.6 eV and the sign of the charge carrier is positive. The shift of the spectra threshold to the shorter wavelength compared with the respective copper phenylacetylenide may be connected with more weak interaction between copper and the ligand due to the sulfur atom. It is remarkable that, as in the case of the metal acetylenides, the photosensitivity was not observed in the CuSR, where R is the bulky substituents. So the steric factors are very substantial for charge transfer. This fact may be confirmed by the dark conductivity. For example the resistivity of the *n*-butyl copper acetylenide is seven orders of magnitude lower than that of the compounds with volume heterocycle ligands.

5.2 Sensitization of the Photoeffect

Copper organoacetylenides were the first compounds in which the chemical and spectral sensitization of the inner photoeffect of the organic materials were realized [19, 20]. The sensitization by dyes adsorbed from the solution was achieved first. The sensitizers tested belong to different dye types, such as the cationic (pinacyanol, methylene blue, rhodamine B, carbo and pentacyanines), the anionic (erythrosin, rose bengal) and the uncharged (chlorophyll, phthalocyanines, hematin). They belong either to photographic sensitizers (pinacyanol) or to desensitizers (methylene blue) [10, 14, 278–282].

Typical results are shown in Fig. 44. The spectral threshold of the proper photoconductivity and the photo-emf of PAC is situated at 520 nm. The spectral response for the photo emf of PAC itself is shown by curve 1. After PAC has been immersed in an ethanol solution of methylene blue and dried its spectral response is represented by curves 2 and 2'. The photo-response appears in the range of the absorption maximum of the dye at 680 nm characteristic of the monomolecular form in the dilute initial solution (curve 3). The observed enhancement of the second maximum at 620 nm in comparison to the solution spectrum is obviously connected with the presence of dye dimers. The shift of the maximum photoresponse to the longer wavelength by 15 nm relatively to the solution is usually the case for the adsorbed state. The sign of the charge carriers both in the proper and sensitized spectra ranges is positive. As seen in Fig. 44 the adsorption of the dye also markedly changes the proper photosensitivity of the PAC. When the monomolecular form of the adsorbed dye dominates, the

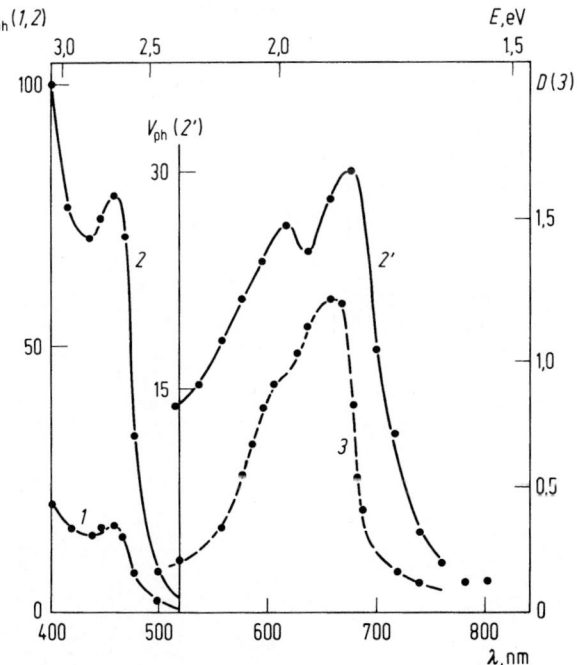

Fig. 44. Spectra for the photoelectromotive force of polycopperphenylacetylenide: curve (*1*) before, curves (*2,2'*) after immersing in 1×10^{-3} ethanol solution of methylene blue; curve (*3*) – absorption spectrum of the solution [20]

positive proper photo-emf is increased by a factor of 5 to 10 without a change in the spectral distribution (curve 2).

It was established that the sign of the photo-emf depends on the time elapsing after dyeing (Fig. 45). Curves 1 and 1′ represent measurements made 10 min after dyeing. The photoresponse spectrum in the sensitized range is different from that of a dilute solution, but similar to that of a solid dye layer. It is remarkable that the sign of the dominant photocarriers in the sensitized spectral range is negative. Moreover, the photo-emf of PAC is now decreased comparatively to the undyed initial sample. On exposure to open air the sensitized photo-emf reverses its sign, becoming positive. Thus the spectral curves 2 and 2′ are for the same sample after 24 hours open to the air. Positive charge carriers appear in the short wavelength range of the sensitized spectrum, whereas in the long wavelength region they are still negative. After prolonged exposure, the sensitized photo-emf becomes positive on the entire spectral range (curve 3, 3′). At the same time an enhancement of the positive photo-emf in the proper spectral range is observed. The reversible cycle with the sample solvent treatment can be performed several times without permanent changes. Apparently such processes reflect the kinetics of the dye adsorption. One can assume the change of the photo-emf sign due to the band bending because of the solvent traces.

The optimum sensitization efficiency from 30 to 50% was observed at concentrations of dyes 10^{-3} M. The experimental data for photoconductivity of PAC sensitized by methylene blue are shown in Fig. 46. It is significant that a

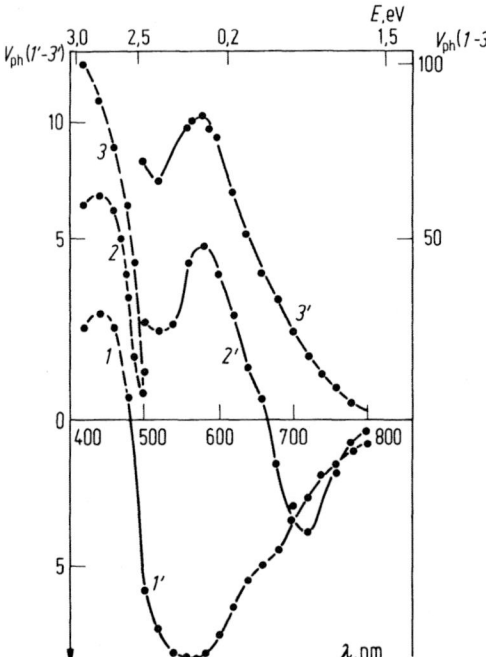

Fig. 45. Photoelectromotive force spectra for polycopperphenylacetylenide sensitized by methylene blue. Curves (1,1′) after 10 min stay in air; curves (2,2′) the same sample after 24 hours stay; curves (3,3′) the same after five days stay [20]

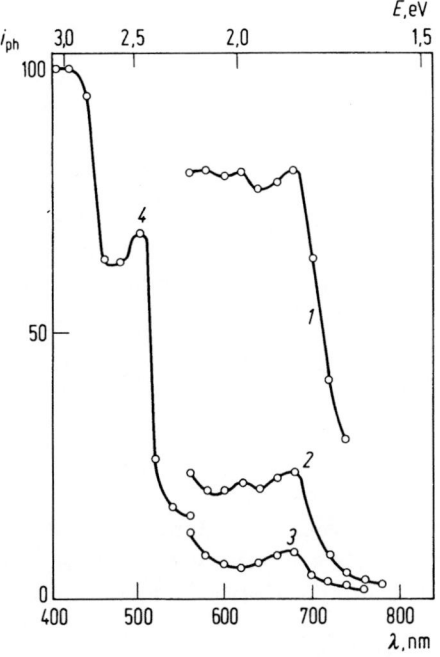

Fig. 46. Photoconductivity spectra of polycop-perphenylacetylenide sensitized with methylene blue ethanol solutions of the following concentrations: $1 - 10^{-3}$ M; $2 - 10^{-4}$ M; $3 - 10^{-2}$ M; 4 – photoconductivity in the proper photosensitivity range of polycopperphenylacetylenide [20]

decrease in the dye concentration is accompanied by an increase in the long wavelength maximum at 680 nm, the sensitization spectrum approaching that of a dilute dye solution. The participation of dimers and more aggregated dye forms in the sensitization appears on increasing the dye concentration.

Sensitized photoconductivity of up to 1300 nm was demonstrated for PAC using various types of dyes [279]. The main features for infrared sensitizers were the same as for visual ones.

As already shown, the sensitized spectra follow the dye absorption. Most of the dyes have sufficiently narrow absorption bands. This does not permit us to obtain the panchromatic sensitivity in the sufficiently broad spectral range. It was proposed to use the polymers with conjugated bonds as sensitizers [21]. The broad diffuse absorption spectra are inherent to such compounds. One can expect higher thermal stability from such sensitizers. In addition the application of binder may be omitted from the preparation of the photosensitive layers, for example, in electrophotography. Polymers with triple bonds, polyphenylenes and polyoxiphenylenes were used as sensitizers [10, 14, 278–280]. The typical results are shown in Fig. 47. The main rules for photoconductivity sensitized by polymers were the same as for the dyes. Optimum sensitization was obtained at the concentration of the sensitizer of 10^{-1}–10^{-2} g/cm^3 relative to the polymeric photoconductor weight.

The analysis of the results obtained on sensitization of the organometallic compounds shows that their principal features coincide with the sensitization of the inorganic semiconductors. Apparently, the energy transfer mechanism from

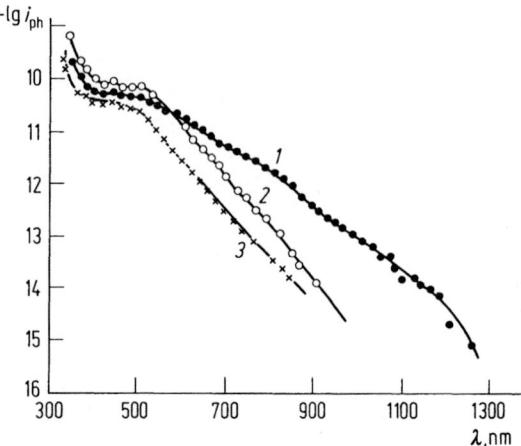

Fig. 47. Photoconductivity spectra of polycopperphenylacetylenide sensitized by polyoxiphenylene with concentrations of the solution in g/cm³ *1* – 10⁻¹, *2* – 10⁻², *3* – 10⁻³ [289]

the dye to the polymer matrix is preferable. The essential argument in favour of such a mechanism is the mutual disposition of the energetic levels for PAC and sensitizers [279] shown in Fig. 48. The levels of the dyes may be both above (pynacyanol, chlorophyll, kryptocyanine) and lower (methylene blue, erythrosin) than the ground PAC state. For hole transfer to the p-type semiconductor it is necessary that the ground singlet level of the dye is lower than the upper part of the valence band. For infrared sensitizers even correction on the adsorbed state cannot guarantee the arrangement of the levels needed for hole transfer. So the mutual disposition of the energetic level of the dye and polymer semiconductors is not the main factor and one can favour the energy transfer mechanism of spectral sensitization.

Besides spectral sensitization the chemical sensitization of PAC was revealed [281–282]. The treatment of the polymer with nucleic acids lead to an increase in the photocurrent by 3 orders of magnitude without changing the photoconductivity spectrum. The same results were obtained with adenine. The data obtained were explained by the model with new recombination centers leading to an increase in the life time of the predominant charge carriers. This was confirmed by kinetic investigations.

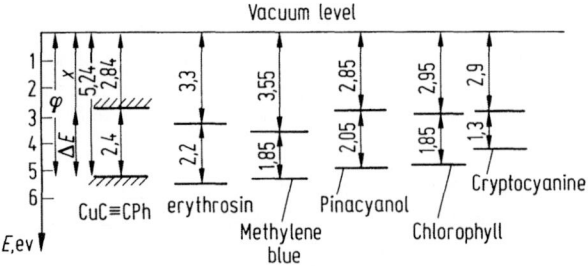

Fig. 48. Scheme of the energetic levels of polycopperphenylacetylenide and dyes [279]. φ – photoelectrical work function, ΔE – the energy of the absorption edge, χ – electron affinity (all data are in eV)

The photosensitivity shown by organometallic compounds permitted their use in electrophotography and optoelectronics [283, 284]. Such materials may also be the bridge connecting the physics of organic and inorganic semi-conductors.

6 Molecule-Doped and Heterogeneous Polymeric Structures

Two types of polymers will be considered in this chapter. Firstly, the polymers with saturated bonds with inserted doping molecules which as a rule do not form charge transfer complexes. These materials may be regarded as the guest-host systems in which the polymer matrix is the host and the doping molecule is the guest. Selective excitation of the doping molecules permits us to obtain detailed information about the photogeneration process. The variation of the concentration of the dopant molecules and the distance between them clears up the role of the disorder and localized states in the charge transfer mechanism. The extreme case of such systems may be the multilayered structures, when one of the component takes part in photogeneration and another in the charge transfer processes.

The second type of system is the group of polymers produced by radiation, thermal, or plasma discharge treatment. The structure of such products is, as a rule, not known due to the high heterogeneity.

6.1 Molecule-Doped Polymers

At the early stages the photoconductivity of solid solutions of the leucobase of malachite green in various organic media was investigated [285]. In these systems, carrier transport occurs by direct interaction between the leucobase molecules. No direct participation of the organic matrix in the charge transfer was observed. A model was proposed which links charge transfer in these systems with impurity conduction in semiconductors.

The photosensitivity spectra of the saturated polymers containing phthalo-cyanine are shown in Fig. 49 [286]. The photosensitivity is due to the presence of the dye. The pure β form magnesium phthalocyanine has nearly the same spectra. Apparently the polymer matrix environment promotes the β-form formation. The long wavelength maximum is related to the aggregated dye form with solvent inclusion.

Later on a lot of research into the photosensitivity of the polymer matrix with various inserted molecules was carried out. A great deal of attention was focused on molecule doped polycarbonates. The photoconduction mechanism in such systems has many common features with the same mechanism for doped PVC.

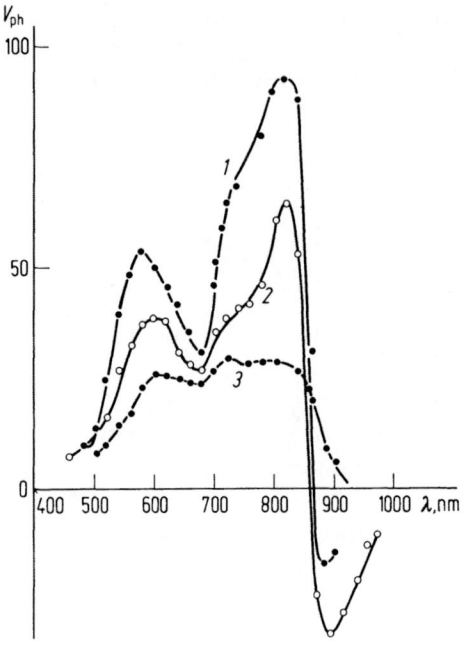

Fig. 49. Photoelectromotive force spectra of the magnesium phthalocyanine in polystyrol (*1*), polyvinyl carbazole (*2*), luvican (*3*) [285]

Usually poly (bisphenol – A carbonate, commercially – lexan) is used as a polymer matrix.

The chemical structure of this polymer is

$$\left(-O-\!\!\left\langle\bigcirc\right\rangle\!\!-\overset{\overset{\displaystyle CH_3}{|}}{\underset{\underset{\displaystyle CH_3}{|}}{C}}-\!\!\left\langle\bigcirc\right\rangle\!\!-O\overset{\overset{\displaystyle O}{\|}}{C}-\right)_n$$

Triphenylamine (TPA) and isopropylcarbazole (IPC) were used at the beginning as a model doping molecules [287–289]. The introduction into the polymer matrix IPC molecules with a concentration of 10^{20} cm^{-3} leads to the clearly seen photogeneration and charge transfer.

Onsager's model was used for explaining the electrophotographic data. For TPA-doped films the thermalization distance of 2.2–2.7 nm and quantum yield of 10^{-2} was obtained. Both values do not depend on the temperature and increase with the increase of the TPA concentration (Fig. 50). It was concluded that the low intrinsic photogeneration is due to both the low primary quantum yield and the high recombination probability of the bound electron-hole pairs. A more detailed explanation in the framework of Onsager's model was obtained for polystyrene and polyvinylbutyrale with TPA. For IPC molecules a stepped photogeneration efficiency increase was observed with a decrease in the wavelength (Fig. 51) [289]. The increase in photogeneration efficiency was explained

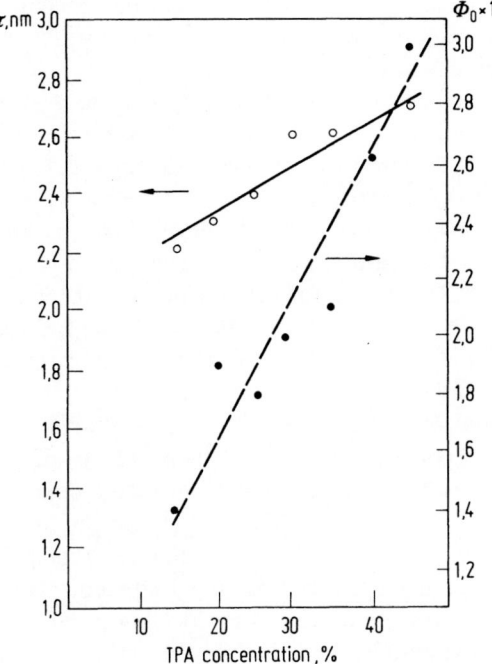

Fig. 50. Dependence of the thermalization distance r and initial quantum yield Φ_0 for polycarbonate film on triphenylamine concentration [287]

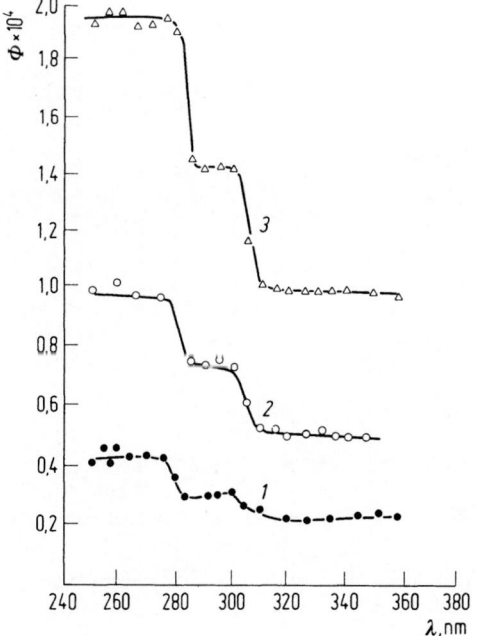

Fig. 51. Spectral dependence of the photogeneration efficiency for polycarbonate film with concentration of isopropylcarbonate 40% [289]. Electric field strength in $V\,cm^{-1}$: $1 - 5 \times 10^4$; $2 - 10^3$; $3 - 1.5 \times 10^3$

by an increase in the thermalization distance and dependence on the concentration of the initial quantum yield. The thermalization distance r did not depend on the temperature and dopant concentration.

The mobility of the charge carriers depends on the temperature, dopant concentration, and electric field strength [11, 14]. For the ideal case, when the size of the local centers with radius R_o is less than the mean distance between them ρ_0, the mobility is given by $\mu \sim \rho^2 \exp(-j\rho) \exp(-\Delta/kT)$, where $j = R_o/2$ is the localization parameter. The exponential dependence of the hole mobility on the distance between the doping molecules was established.

The localization radius R_o was equal to 0.1 nm for lexan with TPA, 0.154 nm with IPC, and 0.1 nm for PVC with TPA. The strong exponential dependence of the mobility proves the charge carrier transfer between localized states, connected with doping molecules.

The drift mobility has the thermal activation energy, which increases with decreasing TFA concentration. The experimental data are shown in Fig. 52 for lexan films with TPA [290]. The drift mobility and its activation energy depends on the electric field strength. For the fields from 20 to 80 $V\mu m^{-1}$, the activation energy decrease is given by $\Delta(E) \sim \Delta_o - \beta E$ [290]. The coefficient increases slightly with the increase in the hopping length, but for fixed concentrations there is apparently a weak dependence on the nature of the doping molecules. The complex properties may be explained by introducing the effective temperature T_{ef} instead of the experimental temperature. For lexan-TPA and

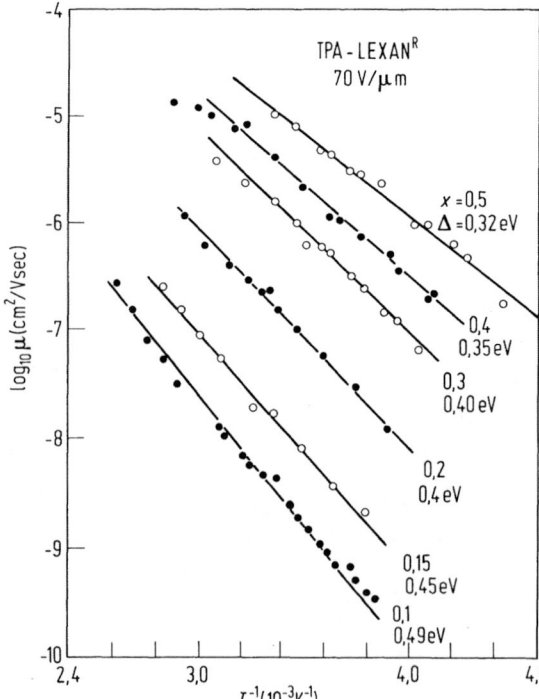

Fig. 52. Hole drift mobility in triphenylamine-Lexan films as function of temperature. $\times$ – mass ratio of the dopant to polymer; Δ – activation energy of the hole mobility (290)

PVC-TPA systems the phenomenological equality $1/T_{ef} - 1/T - 1/T_o$ was established, where T_o – experimental temperature which characterize the transport properties of the system. The molecules securing the weak dependence of the mobility on the electric field may be characterized by low T_o (high T_{ef}).

The strong dependence of the mobility on the mean distance between doping molecules proves the hopping mechanism of the charge transfer. This mechanism may be regarded as an oxidation-reduction reaction. The molecule has to be of the donor type in the neutral state for hole transfer. This takes place for TPA and IPC molecules, so only hole transfer is observed in polycarbonates with these molecules. For electron transfer, the doping molecules have to be of the acceptor type. In lexan with TNF only electron transfer between TNF molecules was established. So the chemical nature of the dopant determines, as a rule, the sign of the charge carriers. In such systems monopolar transfer is realized, instead of CT in PVC-TNF, where bipolar transfer of electrons between TNF acceptor states and holes between chromophore groups of PVC occurs.

For p-type conductivity one can expect that only neutral impurities, whose ionization potential is less than the potential of the molecules actively taking part in the charge transfer, will act as traps. Moreover the impurities with high ionization potential are not essential for charge transfer due to the added activation energy needed for holes to be transported to these localized states often. These proposals were confirmed for lexan films with IPC and TPA [291]. So the influence of the outside molecule on the charge transfer can be predicted knowing the relevant ionization potential and electron affinity.

The dispersive transport processes were discovered in the first molecule doped systems. The experimental data are shown in Fig. 53 for lexan films with

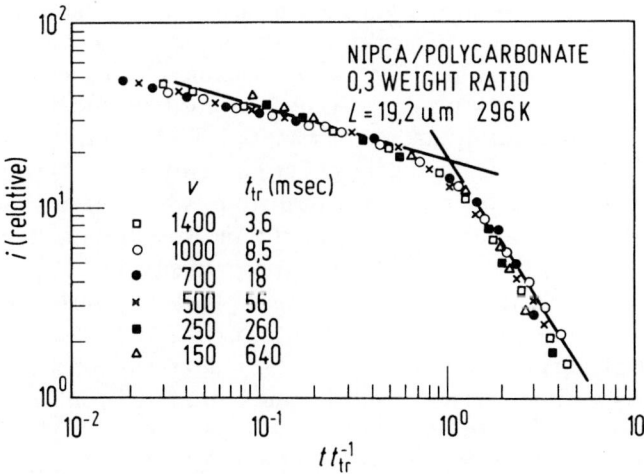

Fig. 53. Master plot of transient hole traces in isopropylcarbazole-Lexan film at room temperature in units log i vs log t/t_{tr}. The traces were normalized to t_{tr} and shifted along the log i axis for the best superposition [292]

IPC [292]. The data are normalized to the transit time and shifted along the current axis for clarity. However, as in the case of CT, there are some difficulties in formulating a simple interpretation of the results obtained due to the impracticability of the Gaussian statistics because the necessary superlinear dependence of the transit time on the sample thickness was not observed.

Carrier photogeneration and transport properties in polycarbonates with phenyl carbazole were also investigated [293–295]. The hole mobility increased from 1.7×10^{-12} to $6.6 \times 10^{-10} \, \mathrm{m^2 \, V^{-1} \, s^{-1}}$ and was observed with the change of the dopant concentrations from 7.5×10^{-4} to $2.4 \times 10^{-3} \, \mathrm{M \, cm^{-3}}$. The micro-brownian motion of the polymer chain was shown to change the energetic trap parameters for hole transport in the glass transition region.

For biscarbazolylcyclobutane-doped polycarbonate films the hole mobility increased drastically with increasing dopant concentration. The hole mobility of $10^{-10} \, \mathrm{m^2 \, V^{-1} \, s^{-1}}$ was analyzed by a random hopping model. The localization radius ~ 1.9 Å was larger that obtained for IPC. This suggests that the larger localization radius is related to the larger spatial extent of the dopant molecules. The low mobility activation energy was ascribed to the absence of an eximer-forming site that works as a multiple-trapping site for hole carriers.

The effective carrier mobilities and their dependence on concentration for benztriazole derivatives embedded in polycarbonate were explained by the percolative aspects in photoconductivity [296]. The observed field dependence of the mobility for polycarbonate films doped with diethynylaminobenzaldehy-de-diphenyl hydrazone cannot be accounted for by any known hopping model [297]. The influence of the nature of the polymer matrix on photogeneration and transport properties of the molecule doped polymers was investigated in some papers [57, 58, 298, 299].

Molecule doped polymers may be successfully used for clarifying the sensit-ization mechanism of the photoconductivity. One can easily change the sensi-tizer concentration and that of the molecules on which the charge carrier transport occurs. Lexan films containing TPA and dye molecules have been used for such purposes [300]. The twin role of the dye, as sensitizer and a trap has been clearly demonstrated for lexan film with fixed TPA content and varying concentrations of Rodamine. Firstly, the photosensitivity increases with increasing dye content. Then the dye acts as a deep trap, which leads to a decrease in the photosensitivity. In contrast to rodamine, the thiapyrylium dyes act as shallow traps, which reduce the charge carrier velocity in hopping transport. The increase in the electric field strength leads to the delocalization of the charge carriers, their transfer and hence to effective sensitization.

The Onsager theory of geminate recombination was qualitatively consistent for aryl-substituted thiapyrylium salt and dialkylamino-substituted triphenyl-methane dispersed in polycarbonate film [301]. The quantum yield of the photogeneration was equal to 0.5 at the electric field strength of $10^6 \, \mathrm{V \, cm^{-1}}$, mobility of $10^{-12} \, \mathrm{m^2 \, V^{-1} \, s^{-1}}$. Hole and electron conductivity was established. In a triphenylamine-lexan system doped with a boron diketone acceptor, the best fit of theory to experimental data was obtained if one assumes a Gaussian distribution of charge transfer radii [302]. A short-range mobility in triphenyl-amine-lexan of the order $10^{-6} \, m^2 \, V^{-1} \, s^{-1}$ was predicted assuming a purely coulombic electron-hole interaction. The most probable CT radius is 0.54 nm with a quantum yield for CT states of 0.93. The model also predicts temperature-dependent quantum efficiency for the doped system more accurately than does the Onsager theory of germinate recombination. Heterogeneous systems containing low molecular organic molecules dispersed in polymer matrix are promising for obtaining high photosensitivity [303–307]. The action spectrum of the photocurrent corresponds to the absorption spectrum of the low molecular compounds. The most interesting results were obtained for phthalocyanines dispersed in polymers [304–306].

It was pointed out earlier that photosensitivity may be realized not only by the inclusion of dyes in a polymer matrix, but also by means of multilayered system production. For TPA dispersed in a polycarbonate, sensitized by a thin layer of vacuum-deposited amorphous selenium, the quantum yield was equal to 0.7 at the electric field strength $6 \times 10^5 \, V \, cm^{-1}$ [308]. The photoinjection efficiency of holes into polymer was equal to the efficiency in pure selenium. The spectra of the quantum efficiency is shown in Fig. 54.

Multilayered systems with organic and inorganic photoconductors and organic transport layers have become the most promising for obtaining high photosensitivity with broad spectral distribution [309–316]. The sensitivity up to $3 \times 10^{-7} \, J \, cm^{-2}$ at the light wavelength 800 nm was obtaining for a layered photoreceptor incorporating phthalocyanines and a hole transport polymer [310, 316].

In addition to heterogeneous systems, various polymers with incorporated inorganic and organic photoconductors are also very interesting for enhancing photoelectrical properties and understanding of the photogeneration, charge transfer, and sensitization mechanisms [79–81]. Such systems are very promising for practical application in various fields of molecular electronics, reproduction processes, holography and so on [14].

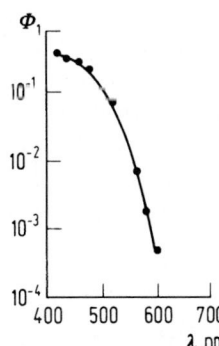

Fig. 54. Spectrum of the photoinjection efficiency from selenium layer into polycarbonate film with 35% content of triphenylamine [308]

6.2 Materials Produced by Thermal, Radiation, and Plasma Polymerization Methods

The polymers produced by thermal, radiation and plasma polymerization methods were among the first organic materials with semiconducting properties [97–99]. References to the early papers on photoconductivity in such materials may be found in the monograph [14].

Thermal treatment of polyacrylonitrile at 200–300 °C leads to the appearance of conjugated bonds and electronic semiconductor properties. The main photoconductivity maximum is situated at 420 nm. An increase of the temperature treatment shifts the photosensitivity to the long wavelengths. The estimated mobilities were from 10^{-7} to 10^{-4} m^2 V^{-1} s^{-1}. The main results obtained proved the sufficiency of the conjugated bonds for the appearance of the semiconductive properties.

The photoconductivity was observed in polyethylene treated by a thermal-radiation method. Intensifying the thermal treatment increases dark and photoconductivity with bathochromic shifts in absorption and photoconductivity spectra due to the appearance of the conjugated bonds. Exponential dependence of the photocurrent on temperature was demonstrated. The optical activation energy of the photoconductivity was less than the thermal activation energy of the conductivity. High energy electron irradiation of high density polyethylene makes donor- and acceptor-type traps in the material with the main photosensitivity in the near infrared region. The volume changes play an important role in the photoresponse formation.

Polypyrrole polymers have a maximum photosensitivity at 610 and 510 nm which increases to 1400 nm after thermal treatment for 24 h at 300 °C [317]. The established thermal activation energy of the photoconductivity excludes the one-step photogeneration process of the charge carriers.

Pyrolyzed polyimide films at 480–530 °C can change the conductivity from 10^{-18} to 10^{-2} S cm^{-1} and mobility from 10^{-11} to 10^{-7} m^{-2} V^{-1} s^{-1} (Fig. 55) [318]. It has been shown that carrier density increases at the initial stage of the pyrolysis and then the increase of the mobility becomes predominant as the pyrolysis progresses. Hopping charge transfer is the main conductive mechanism.

Glow discharge polymerization methods were used to prepare photoconductive films [319–328]. For monomers containing nitrogen, sulfur, selenium and metals after polymerization the conductivity was enhanced by as much as three orders of magnitude with white light excitation [319]. Addition of iodine traces to the discharge increases dark and photoconductivity by 10^3 times. Discrete activation energies of photoconductivity were found 0.05 eV at below room temperature and between 0.05 and 0.35 eV above. An intermediate but non-conducting state was proposed for a possible model of the photoconductivity.

Polymers prepared by glow discharge polymerization of $(CH_3)_4$-M, where M–C, Si, Ge, Sn were photoconductive in the region of 200–350 nm [320]. Good

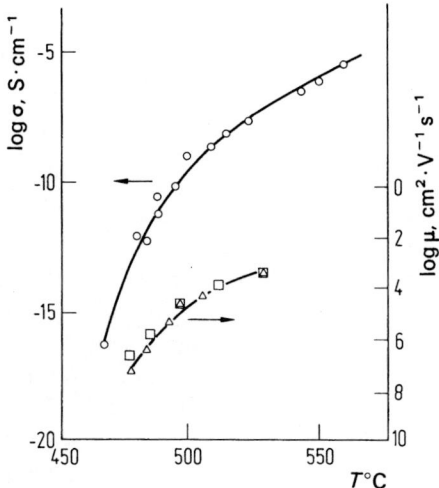

Fig. 55. Hole (□), electron (△) mobilities and conductivity (○) versus pyrolyze temperature for 12 hours of polyimide films [318]

rectification, photovoltaic, photoreduction and electrochromic characteristics at wavelengths of 300–600 nm were demonstrated in plasma-polymerized phthalocyanine films [321]. Photovoltage of 0.9 eV with current density $10^{-7} A \cdot cm^{-2}$ was observed.

Plasma-polymerized phenylacetylene, styrene, PVC, merocyanine dyes and other materials [323–328] were also photoconductive in various spectral ranges. Most of them made Schottky type barriers with electrodes. As a rule, electrophotographic properties attracts attention to polymeric films possessing them.

One can remark that plasma polymerization is one of the most promising methods for obtaining the good quality films with broad range of physicochemical properties.

7 Conclusions. The Status and Prospects of Application of Polymeric Photoconductors

To reiterate the subject of our review it can be easily seen that how much we understand the photoelectrical properties of different types of polymers vary considerably. There are a lot of molecular and supermolecular structures inherent in the materials examined. These permit us to vary the physical and chemical properties over a very broad range and simultaneously lead to great complexity for interpretation of the experimental data in the frame of the simple physical models. The most elegant models have been proposed for relatively simple polymers such as polyacetylene, polydiacetylene, polyvinylcarbazole. However, even here we are at the initial stage of understanding the energetic structure and nature of the conducting and localized states. This understanding must be extended and be adequate for explaining the actual behaviour of more complex polymers. At the same time there is no doubt that chemical structure complication leads to the appearance of properties and possibilities not feasible for simpler compounds.

Photoeffect sensitization in polymers is realized by means of charge-transfer complexes, dyes, polymers, heterogenous and multilayer structures. Every method permits us to obtain some information about photogeneration, charge transfer processes and can be used for concrete practical applications.

In comparison with classical inorganic photoconductors, one can see the low mobilities of the charge carriers $\leq 10^{-5} m^2 V^{-1} s^{-1}$ are usual for the organic polymers. So the most promising applications may be devices where transit distance but not the transit time of the charge carriers is important. Electron acceptor doping permits us to hope that application in both directions may be possible for organic polymers.

Nowadays, polymeric photoconductors may be used in electrophotography, microfilms, photothermoplastic recording, spatial light modulators, and nonlinear elements. The combination of photosensitivity with high quality electrical and mechanical properties permits the use of such materials in optoelectronics, holography, laser recording and information processes. The applications of the various types of polymers were reported in the final parts of the relevant items in the earlier sections. Here, we will briefly analyze the common features of photoconductive polymer applications. The separate questions of each type have been dealt with in some books and papers [3, 11, 14, 329].

Electrophotography. The main stages of the electrophotographic process consist of charging the photoconductive layer on the conducting plate, formation of the latent potential image under illumination, and visualization of the image.

The most commonly used polymers in this field are the ones containing carbazole. Prospective uses may be connected with heterocycle heterogeneous and multilayer systems with sensitivities of up to $10^2 m^2 J^{-1}$ and resolutions $\sim 10^3 mm^{-1}$. The partial transparency permits the use of polymeric photoconductors in microfilms and related fields. The sensitivity of such materials is of the order $1 m^2 J^{-1}$, charge potential 250–400 V, resolution 70–500 mm^{-1}, and transparency 65–70%.

Photothermoplastic recording. A simple variant of photothermoplastic recording consists of the following stages: 1 – charging of the layer in the dark, 2 – exposure with latent electrostatic image formation, 3 – development of the image by thermal treatment. The recording of the information is due to the phase change of the passing or reflected light beam. Molecules of organic polymers guarantee high spatial resolution, technology, dry development and adequate photosensitivity. Single or multilayer structures are now available with sensitivity from the ultraviolet to the near infrared regions. The saturated polymers may be used in the vacuum ultraviolet spectrum and those conjugated or sensitized by CT or dyes and polymers in the visible light region. The sensitized carbazole-containing polymers are usually used as photoconductors in multilayer systems for generation of the charge carriers. A photosensitivity of up to $10^4 \, m^2 \, J^{-1}$ reaching one percent of the diffraction efficiency was obtained. The maximum value for the diffraction efficiency is 20%. Reversible photo-thermoplastic processes may be promising for real time optical processing, holography, image formation, and so on.

Spatiotemporal light modulation. Spatiotemporal light modulators (SLM) are the key elements of optoelectronic systems for optical information processing. The most promising modulators are the liquid crystal type due to the small energy consumption and high parameters. For optical addressing of such SLM, inorganic photoconductors were used until recently. Organic polymeric photo-conductors have been put forward as photosensitive elements in a liquid crystal SLM [254–264]. Low mobilities of the charge carriers permit high spatial resolution. Mechanical and electrical stability, the possibility of sensitization are also factors which make such photosensitive elements in SLM very attractive.

Various types of polymers have been tested in SLM. The best results were obtained with polyimides and polyconjugated polymers. Some of the SLM characteristics are shown in Figs. 31–36. For pulse recording holographic regimes, a limiting resolution up to $1500 \, mm^{-1}$, photosensitivity up to $10^{-8} \, J \, cm^{-2}/\%$, response times $\leq 10^{-4} \, s$ were established. Such parameters are very close to the theoretical limits for thin flat holograms. Using the polymer photoconductor itself for the liquid crystal alignment allows one to simplify SLM production. The information capacity can be easily increased due to the SLM manufacture with higher aperture. An entirely solid state SLM with high parameters has been produced with a photoconductive polymer and polymer dispersed liquid crystal [261]. The SLM, as the phase reversible material, can be used for input and output of the optical information, in optoelectronics, holography, for in-and out-resonance laser light modulation, associative memory, optical interconnections and so on.

Photochemical methods for hologram recording are very similar to the ones mentioned above. Excellent holographic parameters were obtained for poly-diacetylenes [173, 174]. A diffraction efficiency of up to 40% for a spatial frequency of $1600 \, mm^{-1}$ and with a sensitivity of $5 \, cm^2 \, J^{-1}$ was realized. Such a medium may be used in optical processing.

Nonlinear elements. Solar energy conversion is one of the fields undergoing intensive research in order to overcome the energy crisis. Monocrystal silicon batteries have a conversion efficiency of 20 %. The respective efficiencies of 10^{-1}–$10^{-4}\%$ reached by molecular semiconductors seems to be very modest. However, one must remember that the history of molecular batteries is very short. Besides, batteries with high working area may be easily constructed with polymeric photoconductive materials. Economic factors will play the decisive role in manufacturing solar energy conversion devices. Multimolecule structures are promising in this field.

Most photoconductive polymers can be used in solar batteries. The high resistivity of the polymers decreases the actual power of the devices. Possibilities may be connected with electron-donor doping of the polymers. As stated earlier some success has been achieved in this field for polyacetylenes and other conjugated polymers.

For PVC with TNF, polyvinylene, single and multilayer structures the efficiency of the energy conversion was $10^{-2}\%$ [65, 68, 89, 90]. Schottky type barriers with polyacetylenes have a conversion efficiency 0.3–0.01% with fill factor of 0.25. The deficiency of such structures is their instability due to their reaction with oxygen. Heterojunctions of polyacetylene with photoconductors permits us to achieve 0.5% conversion efficiency [135–139]. Photoelectrochemical cells with polyacetylene electrodes had an absolute quantum efficiency at 1% with a photon energy of 2.4 eV [140]. The application of the conjugated polymers in photochemical cells, heterojunctions is also regarded as one of the prospective ways of using polymers as electronic materials.

The diverse properties of organic molecular materials, of which the polymers are amongst the primary ones, will, without doubt, be intensively developed in the future. Photosensitive polymer semiconductors with pre-given properties and a broad spectrum of application will be created for various optoelectronic devices.

8 References

1. Rose A (1963) Concepts in photoconductivity and allied problems. John Wiley, New York
2. Terenin A (1967) Photonics of the dye molecules. Nauka, Leningrad
3. Simon J, Andre J-J (1985) Photoelectrical properties and solar cells. Springer, Berlin Heidelberg New York
4. Pope M, Swenberg C (1982) Electronic processes in organic crystals. Oxford University Press, New York
5. Patsis A, Seanor D (eds) (1976) Photoconductivity in polymers. Konnecticut Technonic Publication
6. Mort D, Pai D (eds) (1976) Photoconductivity and related phenomena. Elsevier, Amsterdam
7. Kryszewski M (1980) Semiconducting Polymers. polish Publishers, Warsaw
8. Mylnikov V (1968) Uspechi Chimii 37: 78
9. Mylnikov V (1974) Uspechi Chimii 43: 1821
10. Mylnikov V (1981) Uspechi Chimii 50: 1872
11. Mort D, Pfister N (eds) (1982) Electronic properties of polymers. Wiley Interscience, New York
12. Meier H (1986) Spectral sensitization. Focal Press, London
13. Akimov I, Cherkasov Y, Cherkashin M (1980) Sensitized photoeffect. Nauka, Moskow
14. Mylnikov V (1990) Photoconductivity in polymers. Chimia, Leningrad.
15. Scher H, Montrol E (1975) Phys Rev 12: 2455
16. Onsager L (1938) Phys Rev 54: 554
17. Gaidelis V (1985) Litvanian Phys Sbornic 25: 87
18. Gaidelis V (1986) Litvanian Phys Sbornic 26: 172
19. Mylnikov V, Terenin A (1964) Molec Phys 8: 387
20. Mylnikov V, Terenin A (1964) Dokl Acad Nauk USSR 155: 1167
21. Mylnikov V, Sidaravichiys I (1968) Electrochimia 4: 596
22. Hogl H (1965) J. Phys Chem 69: 755

23. Borsenberger P, Ateya A (1978) J Appl Phys 49: 4035
24. Kadyrov D, Rumyantsev B, Sokolik I, Frankevich E (1982) Polym. Photochemistry 2: 213
25. Okamoto K, Oda N, Itaya A (1975) Chem Phys Let 35: 483
26. Okamoto K, Itaya A (1984) Bull Chem Soc Jap 57: 1626
27. Kaul H, Haarer D (1987) Ber Bunsenges Phys Chem 91: 845
28. Kuvchinski N, Mostovoj I, Nachodkin H (1985) Ukrain Phys Journal 30: 266
29. Kuvchinski N, Mostovoj I, Pavlov V (1983) Journal Nauchno-Pricladn Photogr Kinemat 28: 465
30. Kuvchinski N, Liasko I, Nachodkin H (1987) Dokl Akad Nauk USSR 294: 1093
31. Stolzerberg F, Ries B, Bassler H (1987) Ber Bunsenges Phys Chem 91: 845
32. Hirsh J, Tahmasli A, (1980) Solid St Com 34: 75
33. Bässler H (1981) Phys Stat Solidi (B) 197: 9
34. Reucroft P, Takahashi K (1975) J. Noncryst Solids 17: 71
35. Fujino M, Mikawa H, Yokoyama M (1982) Phot Science Engin 26: 82
36. Bos F, Burland D (1987) Phys Rev Let 58: 152
37. Tanikawa K, Enomoto T, Hatano M (1975) Macromol Chem 176: 3025
38. Natansohn A, Flaisher H (1984) J Pol Sci Polym Letter Ed 22: 579
39. Itaya A, Okamoto K, Kysabayashi S (1985) Polymer J. 17: 557
40. Giro G, Marco P, Chiellini E (1986) Europ Polym J 22: 801
41. Uryu T, Okamoto H, Oshima R (1987) Macromol 20: 716
42. Undzenas A (1983) Litvanian Phys Sbornic 23: 76, 106
43. Lardon M, Lell-Doller E, Weigl J (1967) Mol Cryst Liq Cryst 2: 241
44. Schaffert R. (1971) J Res Develop (1971) 15: 75
45. Andre B, Lever R, Moisan J (1989) Chem Phys 137: 281
46. Yokoyama M, Endo K, Mikawa M (1981) J Chem Phys 75: 3006
47. Yokayama M, Shimokihara S, Matsubara A, Mikawa H (1982) J Chem Phys 76: 724
48. Yokoyama M, Mikawa H (1982) Phot Sci Engin 26: 143
49. Rumjantsev B (1987) Khimicheska Physica 6: 1352
50. Tsutsumi N, Yamamoto M, Nishijma Y (1988) Polymer 29: 1655
51. Pramanick P, Akhter M (1988) Polymer 29: 752
52. Ikeda M (1991) J Phys Soc Japan 60: 2031
53. Hirohashi R, Kobayashi H, Suzuki T (1990) Polymer Journal 22: 191
54. Gill W (1972) J Appl Phys 43: 5033
55. Qian R, Shi S, Liu T, Zhu D (1989) Phys Stat Sol 111: K67
56. Sasakawa T, Ikeda T, Tazuke S (1989) Macromol 22: 4253
57. Kryukov A, Saidov A, Vannikov A (1989) Khimicheska Physica 8: 1498, ibid (1991) 10: 567
58. Vannikov A, Kruhov Yu, Tyurin A, Zhuravleva T (1989) Phys Stat Sol (a) 115: K47
59. Meier H, Albrecht W, Tschirwitz H (1966) Photochem Photobiol 6: 383
60. Hanna Jun-ichi, Inoue E (1983) Photogr Sci Eng 27: 130
61. Ikeda M, Sato H, Morimoto R (1975) Photogr Sci Eng 19: 60
62. Tani T (1973) Photogr Sci Eng 17: 11
63. Demidov K, Gaziev Z, Akimov I (1983) Journal Nauchno-Pricladn Kinematogr 28: 58; (1985) Journal Nauchno-Pricladn Kinematogr 30: 56
64. Tyurin A, Dubenskov P, Zhuravleva T, Vannikov A (1987) Khimicheska Physica 6: 1236
65. Regensburger P (1968) Photochem Photobiol 8: 429
66. Uss V, Sidaravichjus I (1975) Phys Technica Poluprovodn 9: 1655
67. Rijnnel E (1987) Journal Nauchno-Pricladn Photogr Kinematogr 32: 143
68. Salanek W (1973) Appl Phys Let 22: 11
69. Hague S, Uryu T (1988) Polym J 20: 163
70. Tutihasi S (1972) J Appl Phys 43: 3097
71. Tanikawa K, Okuno Z, Iwaoka T, Hatano M (1977) J Appl Phys 48: 2424
72. Okuno Z, Tanikawa K, Hatano M (1979) Photogr Sci Eng 23: 362
73. Agarwall S, Hemmadi S, Pathak N (1979) Polymer 20: 867
74. Wojcechowski P, Kryszewski M (1979) Acta Physica Polonica A56: 89
75. Kubo I (1988) Jap J Appl Phys 27: 1054
76. Tsutsumi N, Yamamoto M, Nishigima N (1988) Polymer 29: 1655
77. Nakarawa Y, Hoshino K, Jun-ichi Hanna, Kocado H (1989) Jap J Appl Phys 28: 1396
78. Kaneko M, Wöhrle D, Schlettwein D, Schmidt V (1988) Die Macromol Chem 189: 2419
79. Mimura Y, Endo H, Narita K (1973) Jap J Appl Phys 12: 1165
80. Haga Y, Inoue S, Sato T, Yosomiya R (1986) Angew Macromol 139: 49

81. Simuta O, Fukada A, Kuze E (1980) J Polymer Sci Phys Ed 18: 877
82. Wintle H, Tibenski C (1973) J. Polymer Sci A-2: 25
83. Chan G, Wintle H, (1975) J Polym Sci Polym Phys 13: 1187
84. Bune A, Fridkin V, Verkhovskaya T, Taylor G (1990) Polymer J. 22: 7
85. Fridkin V, Shlensky A, Verkhovskaya T (1985) J. Intern Record Materials 13: 421
86. Skotheim T, Inganas O (1985) J. Electrochem Soc 132: 2116
87. Yoneyama H, Wakamoto K, Tamura H (1985) J Electrochem Soc 132: 2414
88. Hagemeister M, Wintle H (1987) J Phys Chem 91: 150
89. Kaneko M, Yamada A, Kenmochi T (1985) J Polym Sci Polym Let Ed 23: 629
90. Kubota E, Yamamoto T (1987) Jap J Appl Phys 26: L1601
91. Nishiguchi N. Uryu T (1988) Polym J 20: 679
92. Hayashida S, Hayashi N (1991) Synth Metal 41: 1243
93. Mylnikov V, Sladkov A, Kudrjavcev Y, Terenin A (1962) Dokl Acad Nauk USSR 144: 840
94. Mylnikov V (1963) Dokl Acad Nauk USSR 148: 120
95. Mylnikov V, Putzeiko E, Terenin A (1963) Dokl Acad Nauk USSR 149: 897
96. Mylnikov V (1964) Dokl Acad Nauk USSR 157: 1184
97. Pohl M, Engelhard E (1962) J Phys Chem 11: 2085
98. Pohl H, Opp D (1962) J Phys Chem 11: 2121
99. Oster G, Oster G, Kryszevski M (1962) J Polym Sci 57: 937
100. Ito T, Shirakawa T, Ikeda S (1974) J Polym Sci 12: 11
101. Epstein J, Conwell E (eds) (1982) Proceed Intern Conf Low Dimentional Conductors Mol Cryst Liq Cryst 77, 79, 81, 83, 85
102. Pecile C, Zerbi G, Bosio R, Girlando R (eds) (1985) Proceed Intern Conf Phys Chem Synth Metals 117
103. Hanack M, Roth S, Schier H (eds) (1991) Intern Conf Synthetic Metals Synth Metals 41, 42, 43
104. Lauchlan L, Etmad S, Chung T-C, Heeger A, McDiarmid A (1981) Phys Rev B243: 3701
105. Etemad S, Heeger A, Lauchlan L (1981) Mol Cryst Liq Cryst 77: 43
106. Etemad S, Heeger A, McDiarmid A (1982) Ann Rev Phys Chem 33: 433
107. Heeger A (1985) Polymer j 17: 101
108. Brasovski S (1978) Pisma Journal Exper Technic Phys 28: 656
109. Brasovski S (1980) Journal Exper Techn Phys 78: 677
110. Brasovski S, Kirova n (1981) Pisma Journal Exper Techn Phys 33: 6
111. Su W, Schrieffer E, Heeger A (1979) Phys Rev Let 42: 1698; (1980) Phys Rev 22: 2099
112. Su W, Schrieffer F (1980) Proc Nat Acad Sci 77: 5626
113. Wegner C (1969) Naturforsch 24: 824
114. Yoshirawa M, Kobayashi T, Fujumoto H, Tanaka J (1987) J Phys Soc Jap 56: 768
115. Kivelson S (1981) Mol Cryst Liq Cryst 65: 79
116. Bredas J-H, Chance P.R, Silbey R (1982) Mol Cryst Liq Cryst 77: 319
117. Bredas J-H, Themans B, Andre J (1983) Phys Rev B27: 7827
118. Frankevich E, Pristupe A, Kobrjansky (1984) Pisma Journal Exper Techn Phys 40: 13
119. Kubar S, Frankevich E (1984) Khimich Physica 3: 964
120. Yakoby Y, Roth S, Menke K, Keilmann F, Kuhl J (1983) Sol St Com 47: 862
121. Bleir H, Roth S, Sheu Y, Schäfer-Siebert H, Leising G (1988) Phys Rev B38: 6031
122. Bleier H Roth S, Lobentanzer H, Leistig G (1987) Europhys Lett 4: 1397
123. Bleier H, Donovan K, Friend R, Roth S, Rothberg L, Tubino R, Vardeny Z, Wilson G (1988) Synthetic metals 88: 204
124. Weinberger B. (1984) Solid State Com 51: 84
125. Kiess H, Keller R (1985) Mol Cryst Liq Cryst 117: 235
126. Grant P, Tani T, Gill W, Street G, Clarke K (1981) Solid State Com 33: 499
127. Kubo T, Sasaki K, Take zoe T, Fukuda A (1987) Jap J Appl Phys 26: 4203
128. Weinberger B, Gau S, Kiss Z (1981) Appl Phys Let 38: 555
129. Tsukamoto I, Ohigashi H, Matsumura K, Takahashi A (1981) Jap J Appl Phys 20: 127
130. Grant P, Tani T, Gill W, Krounk M, Clarke T (1981) J Appl Phys 53: 869
131. Vander Donckt E, Kamicki J, Fedorko P (1984) J Appl Pol Sci 29: 619
132. Kamicki J, Fedorko P (1984) J Phys Appl Phys 17: 805
133. Galluzzi F, Schwarz M (1984) Chem Phys Lett 105: 95
134. Galluzzi F, Schwarz M (1985) Mol Cryst Liq Cryst 121: 301
135. Ozaki M, Peebles D, Weinberger B, Heeger A, McDiarmid A (1979) Appl Phys Lett 35: 83
136. Ozaki M, Peebles D, Weinberger B, Heeger A, McDiarmid A (1980) J Appl Phys 51: 4252
137. Gadene M, Roland M, Abadie J (1983) Revue Phys 18: 691

138. Gadene M, Rolland M, Bougnot J, Abadie J (1985) Mol Cryst Liq Cryst 121: 297
139. Koezuka H, Hyodo K, McDiarmid A (1985) J Appl Phys 58: 1279
140. Chen S, Heeger A, Kiss Z, McDiarmid A (1980) Appl Phys Let 36: 98
141. Berrehar J, Lapersonne-Meyer C, Schott M (1985) Appl Phys Let 48: 630
142. Chance R, Baughmann R, Reueroft R, Takahashi K (1976) Chem Phys 13: 181
143. Chance R, Baughmann R, (1976) J Chem Phys 64: 3889
144. Müller H, Eckhardt C, Chance R, Baughmann R (1977) Chem Phys Let 50: 22
145. Yee K, Chance R (1978) J Pol Sci Phys Ed 16: 431
146. Locher K, Reimer B, Bässler H (1976) Phys Stat Sol 76: 533
147. Reimer B, Bässler H, Hesse J (1976) Phys Stat Sol 73: 709
148. Locher K, Reimer B, Bässler H (1976) Chem Phys 41: 388
149. Reimer B, Bässler H (1978) Phys. Stat Sol (B) 85: 145
150. Locher K, Bässler H, Ticke B, Wegner G (1978) Phys Stat Sol (B) 88: 653
151. Siddiqui A (1980) J Phys C Sol St 13: 1079, 2147
152. Siddiqui A (1984) J Phys C Sol St 17: 683
153. Sieferheld V, Bässler H, Movaghar B (1983) Phys Rev Let 51: 813
154. Sieferheld V, Ries B, Bässler H (1983) J Phys 16: 5189
155. Movaghar B, Cade N (1982) J Phys C Sol St 16: L807
156. Cade N, Movaghar B (1983) J. Phys C Sol St 16: 539
157. Donovan K, Wilson E (1981) Phil Mag 44: 9
158. Movaghar B, Murray D, Donovan K, Wilson E (1984) J Phys C Sol State 17: 1247, 1677
159. Reimer B, Bassler H (1976) Chem Phys Let 43: 181
160. Reimer B, Bassler H (1978) Chem Phys Let 55: 315
161. Blum T, Ries B, Bassler H (1986) J Phys Sol State 19: 3659
162. Donovan K, Wilson E (1985) J Phys Sol State 18: L51
163. Donovan K, Wilson E (1986) J Phys Sol State 19: L357
164. Donovan K, Wilson E (1990) J Phys Sol Cond. Matter 2: 1654
165. Moses D, Sinclair M, Heeger A (1987) Phys Rev Let 58: 2710
166. Moses D, Sinclair M, Philips S, Heeger A (1989) Synth Metals 28: 675
167. Moses D, Heeger A (1989) J. Phys. Cond. Matter 1: 7395
168. Kim Y, Nowak M, Soos Z, Heeger A (1988) J Phys Sol St Phys 21: L503
169. Dubenskov P, Zhuravleva T, Tutnev A, Vannikov A (1987) Khimicheska Physica 6: 764
170. Frankevich E (1991) J Phys Cond Matter 3: 3841
171. Frankevich E, Sokolik J, Lymarev A (1989) Mol Cryst Liq Cryst 175: 41
172. Donovan K, Freeman F, Wilson E (1985) J. Phys Sol State 18: L275
173. Gutter W, Bauer H, Kohler B, Schwoerer (1986) Mol Cryst Liq Cryst 137: 117
174. Bauer H, Vogtmann T, Muller I, Schwoerer (1989) Chem Phys 113: 303
175. Tani H, Tanaka S, Toda F (1963) Bull Chem Soc J 36: 1267
176. Sidaravichjus I, Levina F, Ribalko T, Mylnikov V, Uchin L (1966) Optico-Mechanich Promischl 5: 27
177. Kargin V (ed) (1968) Organic semiconductors. Nauka Moskow
178. Kang E, Ehrlich P, Bhatt A, Anderson W (1982) Appl Phys Let 41: 1136
179. Kang E, Ehrlich P, Bhatt A, Anderson W (1984) Macromolecules 17: 1020
180. Kadyrov d, Kozlov L, Sokolik I, Frankevich E (1983) Khimiya Visoc Energii 17: 68
181. Peleger J, Nespurek S, Vohlidal J. (1989) Mol Cryst Liq Cryst 166: 143
182. Zhao J, Yang M, Shen Z (1991) Pol Journal 23: 963
183. Kang E, Neoh K, Masuda T, Higashimura T, Yamamoto M (1989) Polymer 30: 1328
184. Kashirski J, Sinitzki V, Mijuchina G, Ermakova T, Lopyrev V (1981) Visokomol Soedinen 1: 207
185. Swiatek J (1986) Mol Cryst Liq Cryst 137: 131
186. Tabata M, Satoh M, Kaneto K, Yoshino K (1986) J Phys Soc Jap 55: 1305
187. Beck F (1964) Berichte Bunseges Phys Chem 68: 558, 901
188. Sinitzki V, Mijuchina G, Krijarhin Y, Rosenstein L (1972) Izvest Acad Nauk USSR Khimich ser 4: 969
189. Kaneto K, Uesugi F, Yoshino K (1987) J Phys Soc Jap 56: 3703
190. Kaneto K, Uesugi F, Yoshino K (1987) Sol State Com 64: 1195
191. Vardeny Z, Ehrenfreund E, Brafman O (1986) Phys Rev Let 56: 671
192. Rabe J, Schmidt W, Yoshino K (1985) Jap J Appl Phys 24: L583
193. Yoshino K, Sawada K, Ohada K (1989) Jap J. Appl Phys 28: 1029
194. Kaneto K, Ishi C, Yoshino K (1985) Jap J Appl Phys 24: L320

195. Kaneto K, Takeda S, Yoshino K (1985) Jap J Appl Phys 24: L553
196. Frank A, Glenis S, Nelson A (1989) J Phys Chem 93: 3818
197. Takai V, Inoue M, Shibata A, Tizutami T, Ieda M (1984) Jap J Appl Phys 23: 1614
198. Yoshino K, Mun Soo Yun, Ozaki M, Inushi Y (1984) Jap J Appl Phys 23: L55
199. Mun Soo Yun, Yoshino K (1985) J Appl Phys 58: 1950
200. Tsutsui T, Nitta N, Saito S (1985) J Appl Phys 57: 5367
201. Saito S, Tsutsui T, Tokito S, Hara T, Hsien-Tang Chiu (1985) Polymer J 17: 209
202. Tameev A, Zhuravleva T, Vannikov A, Sergeev V, Nedelkin V, Arnautov S (1985) Dokl Acad Nauk USSR 280: 1398
203. Tameev A, Zhuravleva T, Vannikov A, Sergeev V, Nedelkin V, Arnautov S (1986) Khimicheska Physica 5: 106
204. Tameev A, Zhuravleva T, Vannikov A, Sergeev V, Nedelkin V, Arnautov S (1987) Visokomol Soedinen 29: 2186
205. Tanabe Y, Shimizu H, Minami N (1988) Jap J Appl Phys 27: 1748
206. Takimoto A, Tanaka E, Watanabe M (1989) Jap J Appl Phys 28: L1252
207. Hörhold H, Opfermann J (1970) Macromol Chem 131: 105
208. Hörhold H, (1972) Zt Chemie 12: 41, 750
209. Sahena A, Gunton J (1987) Phys Rev B35: 3914
210. Friend R, Bradly D, Townsend P (1987) J Phys Appl Phys 20: 1367
211. Bradley D, Schen Y, Bleir H, Roth S (1988) J Phys Sol State 21: L515
212. Yoshino K, Takiguchi T, Hayashi S, Dac Hec Park, Sugimoto R (1986) Jap J Appl Phys 25: 881
213. Tokito T, Tsutsui T, Tanaka R, Saito S (1986) Jap J Appl Phys 25: L680
214. Tokito S, Saito S, Tanaka R, (1986) Macromol Chem Rapid Com 7: 557
215. Kryukov A, Vannikov V, Pachredinov A, Horhold H, Ophermann J (1990) Visocomol Soedinen 32: 348
216. Gailberger M, Greiner A, Bässler H (1991) Synth Met 41: 1269
217. Pentch S, Yang J-P, Li H-L (1991) Synth Met 41: 1369
218. Takai Y, Ishii K, Mizutani T, Ieda M (1978) J Phys Appl Phys 12: L139
219. Takai Y, Ishii K, Mizutani T, Ieda M (1979) J Phys Appl Phys 12: 601
220. Takai Y, Kurachi A, Mizutani T, Ieda M, Seki K, Inokuchi H (1982) J Phys Appl Phys 15: 917
221. Mori T, Mizutani T, Ieda M (1990) J Phys Appl Phys 23: 338
222. Voischev V (1977) Electrical properties of the polyheteroarylenes Thesis, Institute of Visokomolecul Soedineniy, Leningrad
223. Pravednikov A, Kotov B, Pebalk D (1987) In: Kolotyrkin (ed) Physicheska Khimij. Khimia, Moskow, p.165
224. Pillai P, Sharma B (1979) Polymer 20: 1431
225. Sharma B, Pillai P (1982) Phys Stat Sol 71: 583
226. Takai Y, Mal-Mun Kim, Kurachi A, Mizutani M, Ieda M (1982) Jap J Appl Phys 21: 1524
227. Quamara J, Bhardway R, Sharma B (1984) Appl Phys A35: 267
228. Rashimi S, Yoshiaki T, Mizutani T, Ieda M (1985) Jap J Appl Phys 24: 1003
229. Iida K, Waki M, Nakamura S, Ieda M, Sawa G (1984) Jap J Appl Phys 23: 1573
230. Iida K, Tanimoto T, Nakamura S, Ieda M (1986) Jap J Appl Phys 25: 1542
231. Iida K, Nohara T, Nakamura S, Sawa G (1989) Jap J Appl Phys 28: 1390, 2552
232. Sessler G, Habur B, Yoon D (1986) Appl Phys 60: 318
233. Parizer R (1987) Polym J 19: 127
234. Freilich S (1987) Macromol 20: 973
235. Vasilenko N, Kotov B, Rybalko G, Kaplunova L, Gaidelis V, Sidaravichjus I, Pravednikov A (1983) Journal Nauchn Pricladn Photogr Kinematogr 28: 349
236. Dubenskov P, Zhuravleva T, Vannikov A, Vasilenko N, Berendiaev V (1988) Visocomolec Soedinen 30: 1211
237. Loutfy R, Hor A, Kazmaier P, Tani M (1989) J Imag Science 33: 151
238. Takimoto A, Wakemoto H, Ogawa H (1991) J Appl Phys 70: 2799
239. Nespurek S (1973) Check J Phys 23: 368
240. Plulips S, Yu G, Cao Y, Heeger A (1989) Phys Rev B39: 10702
241. Monkman A, Bloor D, Stevens G, Stevens J (1987) J Phys D Appl Phys 20: 1337
242. Sergeev V, Nedelkin V, Timopheeva G, Bachmutov V, Yupherov A, Turin A, Zhuravleva T, Vannikov A, (1987) Visokomolec Soedinen 29: 1638
243. Pramanick P, Akhter M (1991) Polymer 32: 160
244. Marko P, Giro G, Lore S (1985) Mol Cryst Liq Cryst 118: 439
245. Musser M, Dahlberg S (1980) J Chem Phys 72: 4048

246. Kosmel G, Bocionek P (1981) Macromol Chem 182: 3445
247. Meier H (1985) Synth Metals 11: 333
248. Belaish J, Rettori G, Davydov A, Yu L, Melean R, Dalton L (1989) Synth Metals 33: 341
249. Ulanski J, Sielski J, Glatzhofer D, Kryszewski M (1990) J Phys Appl Phys 23: 75
250. Turin A, Kryukov A, Zhuravleva T, Vannikov A (1988) Khimicheska Physika 7: 1703
251. Meier H, Albrecht W, Risch H, Nusslan F (1989) J Phys Chem 93: 7726
252. Ohkawa H, Fuznichi T, Oshima R, Uryu T (1989) Macromolecules 22: 2266
253. Stöckert D, Kessel R, Schultze J (1991) Synth Metals 41: 1295
254. Mylnikov V, Morozova E, Vasilenko N, Kotov B (1985) Dokl Acad Nauk USSR 281: 897
255. Mylnikov V, Morozova E, Vasilenko N, Kotov B, Pravednikov A (1985) Journal Technich
     Phys 55: 749
256. Mylnikov V, Groznov M, Morozova E, Soms L, Vasilenko N, Kotov B (1985) Pisma Journal
     Technich Phys 11: 38
257. Mylnikov V, Groznov M, Soms L, Tarasov A (1987) Journal Technich Phys 57: 2041
258. Mylnikov V, Ivanov A (1987) Journal Technich Phys 57: 598, 729
259. Mylnikov V, (1987) Mol Cryst Liq Cryst 152: 597
260. Adomenas P, Vasilenko H, Groznov M, Mylnikov V, Slusar A, Soms L, Schwetz V (1991)
     Journal Technich Phys 61: 185
261. Groznov M, Vasilenko N, Mylnikov V, Mokshin V, Topol C, Truchtanov V, Schwetz V, Soms
     L (1991) Journal Technich Phys 61: 80
262. Mylnikov V, Slusar A (1991) Synth Metals 43: 1341
263. Mylnikov V, Slusar A (1991) Journal Technich Phys 61: 347
264. Mylnikov V, Slusar A (1991) in: Books of Abstracts IY Internat Meeting on Optics of Liquid
     Crystals Cocoa Beach, Florida, p.76
265. Mylnikov V, Terenin A, (1963) Dokl Acad Nauk USSR 153: 1381
266. Konovalov L, Mylnikov V (1974) Journal Strukt Khimii 15: 709
267. Mylnikov V (1974) Journal Strukt Khimii 15: 244
268. Denisov E, Smirnov K, Mylnikov V (1974) Physica Tverdogo Tela 16: 1782
269. Suchov D, Mylnikov V (1975) Physica Tverdogo Tela 17: 923
270. Laschkov G, Mylnikov V (1986) Optica and Spectroskopy 20: 86
271. Mylnikov V, Dunje A (1971) Optica and Spectroskopy 31: 405
272. Mylnikov V, Dunje A, Golding J (1972) Journal Obcshei Khimii 42: 2543
273. Mylnikov V, Golding I, Sladkov A (1976) Izvest Acad Nauk USSR, ser Khim 2: 330
274. Mylnikov V, Golding I, Sladkov A (1978) Izvest Acad Nauk USSR, ser Khim 4: 823
275. Mylnikov V (1965) Dokl Acad Nauk USSR 164: 622
276. Mylnikov V (1976) Journal Physich Khimii 5: 1979
277. Mylnikov V, Charchenko A, Sobolev M (1976) Quantovoya Electronica 3: 288
278. Mylnikov V, Terenin A (1968) J Polym Sci 16: 3655
279. Mylnikov V (1979) Journal Nauchn Prikladn Photogr Kinematogr 24: 25
280. Mylnikov V, Dunje A, Nizova S (1971) Journal Nauchn Pricladn Photogr Kinematogr 2: 132
281. Mylnikov V (1967) Dokl Acad Nauk USSR 175: 726
282. Mylnikov v (1970) J Polym Sci 30: 673
283. Sidarvichjus I, Levina P, Mylnikov V (1966) Optico-Mechanich Promischlen 5: 27
284. Sidarvichjus I (1969) J Appl Optics 30: 779
285. Mehl W, Wolf N (1964) J Phys Chem Solids 25: 1221
286. Mylnikov V (1968) Journal Physich Khimii 42: 2168
287. Borsenberger P, Contois G, Hoestery D (1978) J Chem Phys 68: 637
288. Borsenberger P, Contois G, Hoestery D (1978) Chem Phys Let 56: 574
289. Borsenberger P, Contois G, Ateya A (1979) J Appl Phys 50: 914
290. Pfister G (1977) Phys Rev B16: 3676
291. Pfister G, Mort G, Grammatica S (1976) Phys Rev Let 37: 1360
292. Mort J, Pfister G, Grammatica S (1976) Solid State Com 18: 693
293. Tsutsumi N, Yamamoto M, Nishijima Y (1986) J Appl Phys 59: 1557
294. Tsutsumi N, Yamamoto M, Yasunori A (1987) J Polym Sci Pol Phys 25: 2139
295. Tsutsumi N, Yamamoto M, Nishijima N (1988) Polymer 29: 1655
296. Domes A, Leyrer R, Haarer D, Blumen A (1987) Phys Rev B36: 4522
297. Schein L, Rosenberg A, Rice S (1987) J Appl Phys 60: 4287
298. Kanemitsu Y, Imanishi D, Imamura D (1989) J Appl Phys 66: 4526
299. Turin A, Krukov Yu. Zhuravleva T, Vannikov A, Markevich N, Post M (1988) Journal Nauchn
     Pricladn Photogr Kinematogr 33: 418

300. Grammatica S, Mort J (1977) J Chem Phys 67: 5628
301. Borsenberger P, Growdy A, Hoestery D (1978) J Appl Phys 49: 5555
302. Goliber T, Perlstein J (1984) J Chem Phys 80: 4162
303. Uss S, Uss V, Undzenas A, Sidaravichjus I (1986) Journal Nauchn Pricladn Photogr Kinematogr 31: 438
304. Kaneko M, Wohrle D, Schlettwein D, Schmidt V (1988) Die Macromol Chemie 189: 2419
305. Abdel-Malik T, Ahmed A, Riad A (1990) Phys Stat Sol 121: 507
306. Nakamura S, Ozaki T, Toviyama K, Iida K, Sawa G (1989) Jap J Appl Phys 28: 991
307. Mack J, Shchein L, Peled A (1989) Phys Rev B39: 7500
308. Mey W, Borsenberger P, Crowdry F (1978) J Appl Phys 49: 3607
309. Shimidzu T, Toda T (1987) Pol J 16: 919
310. Loutfy R, Hor A, Pucklidge A (1987) J Imag Sci 31: 31
311. Sama M, Kennell E, Braun C (1987) J Chem Phys 87: 6766
312. Kanemitsu Y, Imamura S (1989) Appl Phys Let 54: 872
313. Kanemitsu Y, Imamura S (1990) J. Appl Phys 67: 3728
314. Kanemitsu Y, Funada H (1991) J Phys Appl Phys 24: 1409
315. Umeda M, Nimi T, Hashimoto M (1990) Jap J Appl Phys 29: 2746
316. Enokida T, Hirohashi R, Kurata M (1991) J Appl Phys 70: 3942
317. Reucroft P, Scott H, Serafin F (1968) J Polym Sci Part C 30: 261
318. Hara T, Tsutsui T, Saito S (1985) Jap J Appl Phys 24: 970
319. Bradley A, Hammes J (1965) Trans Farad Soc 61: 773
320. Inagaki N, Mitsuuchi K (1984) J Polym Sci Let 22: 6301
321. Osada Y, Mizumoto A (1986) J. Appl Phys 59: 1176
322. Phadke S, Kanekap R (1978) Phys State Solid 45: K163
323. Guastarino J, Carchano H, Bui A (1975) Thin Sol Films 27: 225
324. Morita S, Shen M (1977) J Pol Sci Phys 15: 981
325. Sawa G, Murata T, Ohara M, Sunda F, Nakamura S (1986) Jap J Appl Phys 25: 53
326. Inoue M, Morita H, Takai Y, Mizutani T, Ieda M (1986) Jap J Appl Phys 25: 1495
327. Inoue M, Takai Y, Mizutani T, Ieda M (1986) Jap J Appl Phys 25: 1174
328. Takai Y, Mizutani T, Ieda M (1987) Jap J Appl Phys 26: 812
329. Skotheim T (ed) (1986) Handbook of conducting polymers. New York

Editor: Prof. D. Haarer
Received June, 1992

Synthesis and Evaluation of Metal-Containing Polymers

Mukul Biswas and Amit Mukherjee
Department of Chemistry, Indian Institute of Technology,
Kharagpur 721 302, India

The review aims at highlighting the significant developments during the last decade (1980–1991) in the field of metal-containing polymers with special reference to synthetic methodologies used and evaluation of various properties. The synthetic procedures so far adopted include (a) polymerization/copolymerization of metal-complexed monomer moieties (b) anchoring of metal complexes on preformed polymers (c) plasma polymerization (d) doping and (e) mechanochemical synthesis. The variety of monomer/polymer systems studied under the above procedures include the acrylics, styrene derivatives heterocyclics and miscellaneous polycondensates such as polyimides, polyamide-imides, polyethylene oxides, metal porphorazines, phthalocyanines, crown-ethers, Schiff's bases and liquid crystal polymer systems.

The review also highlights, how and to what extent metal ion incorporation can modify the essential bulk properties of the base polymer – such as thermal stability, dielectric, conductivity and other physicochemical characteristics.

In general, the thermal stability of metal-containing polymer systems is relatively enchanced compared to that of the bulk polymer. Various factors including size and concentration of the metal ions, and crystal field stabilization energy of the anchored metal complexes influence the stability to different extents.

The conductivity of many metal modified systems is reportedly enhanced due to various factors such as charge transfer between metal ions and the electron-rich heteroatoms, elimination of impurities, and changes in the transport number of cations and anions due to environmental changes in the solid electrolytes. Even interesting cases have been reported where a polymer film can reach the electronically conducting metallic level by *cis-trans* isomerization.

Metal containing polymers have emerged as a new generation material with tremendous potential in fields like superconducting materials, ultra-high strength materials, liquid crystals and biocompatible polymers.

Advances in Polymer Science, Vol. 115
© Springer Verlag Berlin Heidelberg 1994

1 Introduction

During recent years increasing importance is being laid on research into the synthesis and evaluation of metal-containing polymers. The research on metal-containing polymers started sporadically in the 1950s when Arimato and Haven [1] found vinyl ferrocene could be radically polymerized. The slow progress that continued from the 1960s gathered momentum gradually as evident from the frequent organization of national and international symposia in 1970, 1977–79, 1981, 1983, 1989 and 1992 which saw the field mature [2]. This article is intended to highlight the significant developments in the field of metal-containing polymers during the last decade (1980–1991). The scope of this article will be restricted to the synthetic methodologies adopted and modification of physico-chemical properties of the various polymer systems with the incorporation of metal moieties in the skeleton. The procedures that have usually been adopted for introducing metal ions in the polymers may be conveniently discussed under the following headings: (1) Polymerization/copolymerization of metal-complexed monomeric moieties; (2) Anchoring of metal complexes on pre-formed polymers; (3) Plasma polymerization; (4) Doping and (5) Mechanochemical synthesis. The application of these procedures will be discussed on the basis of selective examples from the various systems explored so far.

2 Synthetic Methodologies

2.1 Polymerization and Copolymerization of Metal-Complexed Monomeric Moieties

This procedure has been used in a limited number of systems and suffers from the obvious disadvantage of the experimental complications involved. Usually, the metal ion is incorporated in the monomeric moiety followed by its polymerization.

2.1.1 Acrylic Monomer Systems

Monomers like iron (III) methacrylate $[CH_2=C(CH_3)-COO]_3Fe$ and tributyltin methacrylate have been synthesized [3]. They undergo free radical polymerization/copolymerization with other acrylates such as cyclohexyl, 2-ethoxyethyl, 2-hydroxyethyl and 2-hydroxypropyl. Synthesis of a metallo-macrocyclic acrylic monomer is illustrated by the reaction between a metallophthalocyanine tetracarbonyl chloride (1) and 2-hydroxyethyl methacrylate (2-HEMA) [4]

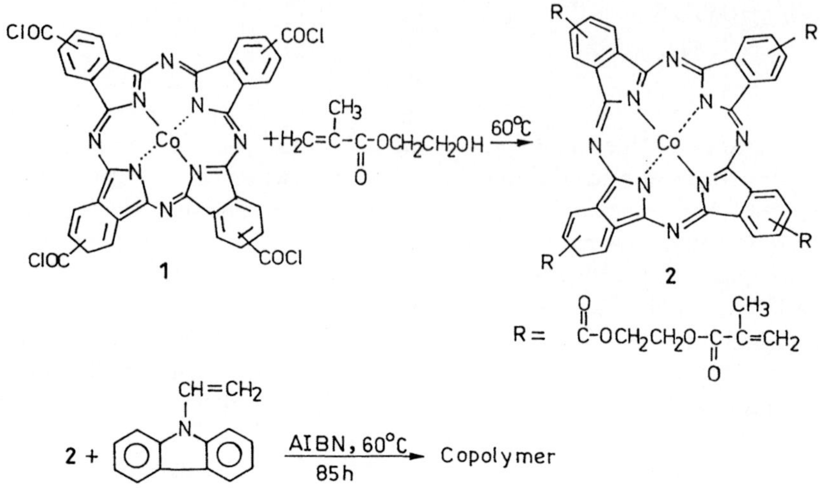

Fig. 1. Metal containing acrylic monomer

(Fig. 1). Copolymerization of this monomer with *N*-vinylcarbazole (AIBN, benzene) results in a blue-green product soluble in usual solvents [4].

2.1.2 Styrene Monomer Systems

A novel procedure [5] is exemplified in the preparation of polystyryl aluminium derivatives by thermal polymerization of styrene in the presence of $AlEt_3$ acting as chain transfer agent.

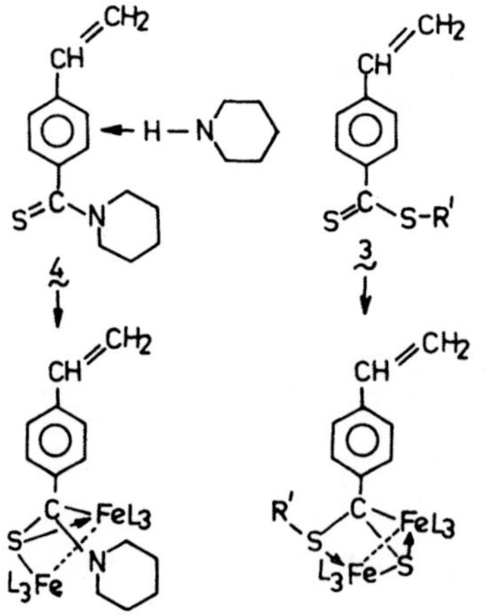

Fig. 2. Metal containing styrene monomer

$$(Polyst)^\bullet + AlEt_3 \rightarrow Polyst. AlEt_2 + Et^\bullet$$

$$Et^\bullet + St \rightarrow Et\text{--}St^\bullet + mSt \rightarrow (Poly\ St)^\bullet$$

$$Poly\ St\text{--}AlEt_2 + (Poly\ St)^\bullet \rightarrow (Poly\ St)_2\ AlEt + Et^\bullet$$

Synthesis of a styrene monomer containing a diiron hexacarbonyl moiety and its copolymer together with the metal atom of the preferred copolymers has been achieved [6]. (3) and (4) (Fig. 2) undergo free radical copolymerization with

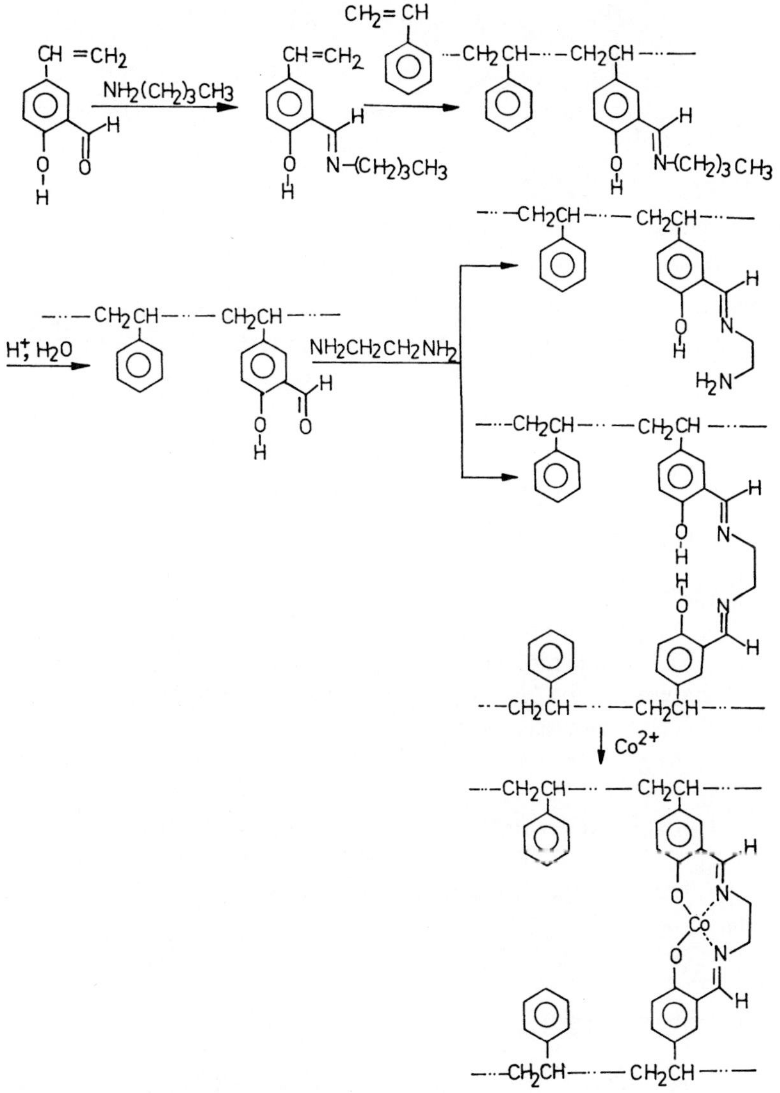

Fig. 3. Cobalt-chelated styrene monomer

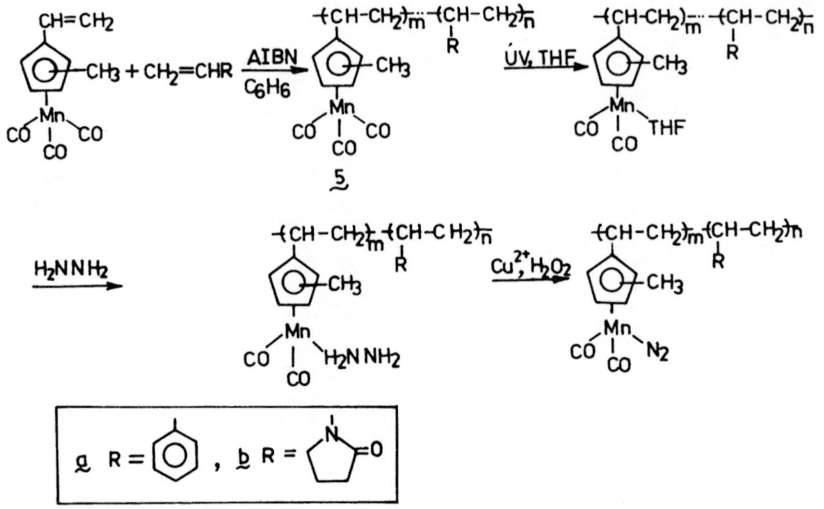

Fig. 4. Mn-bound dinitrogen complex of styrene

styrene, MMA or N-[tris-hydroxymethyl]methacrylamide to afford organo-metallic polymers. However, 3, 4 fail to homopolymerize under free radical conditions.

Synthesis of a typical cobalt chelated styrene derivative [7] is illustrated in Fig. 3.

2.1.3 Miscellaneous Monomers

Polymer-bound dinitrogen complexes containing 'Mn' have been prepared [8] directly from the polymer bound (η^5-vinylmethyl-cyclopentadienyl) tricar-bonylmanganese (VCM) and molecular nitrogen in THF/benzene. The metal-containing base polymer may be readily prepared by AIBN initiated radical polymerization of VCM with styrene or N-vinylpyrrolidone (Fig. 4).

2.2 Anchoring of Metal Complexes on Preformed Polymers

This appears to be the most widely adopted procedure for preparation of metal-containing polymer systems.

2.2.1 Acrylic Polymer Systems

A Co(III)-Chelate [9], bis-2-(2-azopyridyl)-1-naphthol Co(III) has been inter-acted with polyacrylic acid and other polyelectrolytes (Fig. 5).

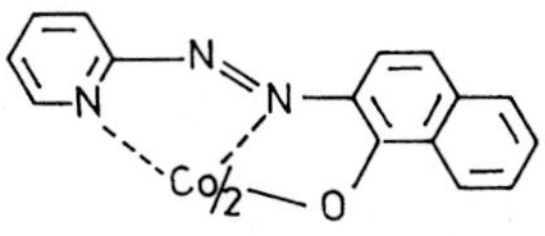

Fig. 5. Co(III)-chelate

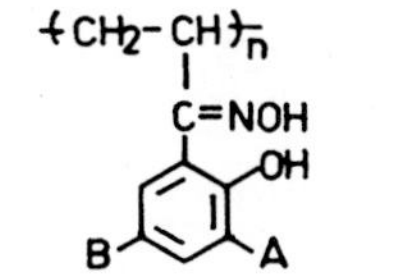

A, B = Substituents

Fig. 6. Ion exchange resin bearing hydroxyloxime group

Metal specific ion-exchange resins bearing hydroxyoxime groups have been prepared [10] from substituted phenylacrylates (Fig. 6).

Silica-supported cyclized polyacrylonitrile (PAN) metal complexes, claimed to be better oxidation catalysts than the polyphthalocyanine-metal complexes, have been obtained shown below as [11]:

| CH$_2$=CHCN supported on silica gel | AIBN; 65 °C (Butanol) $\longrightarrow$ | Silica supported Polyacrylonitrile cyclized at different temperatures (190–250 °C) | Metal Chloride $\longrightarrow$ Cu(I), Cu(II), Co(II), Mn(II) ethanol; reflux | PAN-M-Complex |

By an essentially similar procedure, silica-supported poly-(methacrylic acid)-palladium and platinum complexes have been obtained [12].

Copolymers of methacrylic acid and ethylene termed as 'ethylene ionomers' have been used as the base polymer for binding alkali, alkaline earth and transition metal ions. Organic amines such as *n*-hexylamine, hexamethylene tetraamine, 2,2,6,6-tetramethyl-4-hydroxy piperazine, ethylene diamine and polymeric diamines such as silicone diamine, polyether diamine and polymeric diamines such as silicone diamine, polyether diamine and polyamide oligomers considerably enhance the complex formation characteristics of Zn(II) ethylene ionomers thereby enhancing the physico-chemical properties [13].

Biswas and Moitra synthesized PVC bound dimethylglyoxime (DMG) complexes of Co(II), Ni(II), Cu(II) and Fe(III) by Cl-displacement between PVC and DMG followed by complexation with 3d metal salts [2].

2.2.2 Styrene Polymer Systems

Polystyrene and its divinylbenzene cross-linked copolymer have been most widely exploited as the polymer support for anchoring metal complexes. A large variety of ligands containing N, P or S have been anchored on the polystyrene-divinylbenzene matrix either by the bromination-lithiation pathway or by direct interaction of the ligand with Cl-, Br- or CN-methylated polystyrene-divinyl-benzene network [14] (Fig. 7).

In a novel procedure, chemical modification of uncrosslinked atactic poly-styrene by acetylation followed by a Claisen condensation with ethyl perfluoro-propanoate has been used to prepare a macromolecular ligand bearing phenyl,

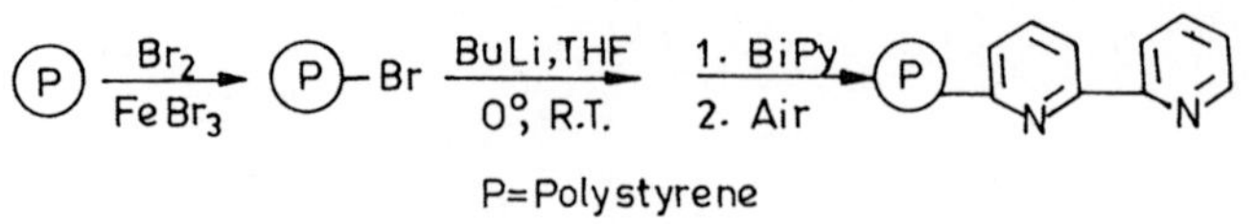

P = Polystyrene

Fig. 7. Lithiation-bromination pathway

perfluoroethyl, 1,3-diketone chelating groups, which readily form chelates with Cu(II), Ni(II) and Uranyl ions [15].

The polystyrene-pendant $Ru(bpy)_3^{2+}$ complex has been likewise obtained by the reaction of dichloro-bis(2,2'-bipyridyl-Ru(II)) complex with bipyridylated polystyrene [16] (Fig. 8).

Coloured complexes of a number of $3d$ metal salts with bipyridylamine previously anchored on partially chloromethylated polystyrene-divinylbenzene matrix have been reported by Hendricker and Kratz [17] and by Biswas and Mukherjee [18] (Fig. 9).

A synthesis of much relevance is that of the copolymer (4-methyl, 4'-vinyl 2,2' bipyridine/styrene)-pendant tris(2,2'-bipyridyl) ruthenium(II) complex as sensitizer for water photolysis with solar irradiation [19].

2-Carboxylbenzoyl and 1-carboxyl 8-naphthoyl partially substituted polystyrene have been prepared by the Friedel Crafts reaction between polystyrene and the corresponding dicarboxylic anhydrides. Rare earth (Europium-III) complexes of these polymer based ligands have been obtained [20] (Fig. 10).

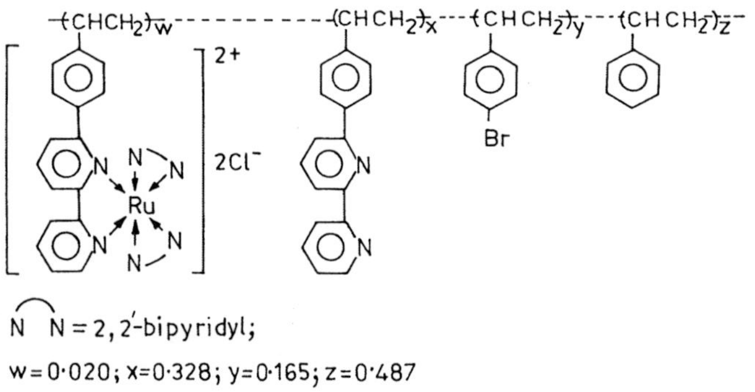

$N \frown N = 2,2'$-bipyridyl;

$w = 0.020; x = 0.328; y = 0.165; z = 0.487$

Fig. 8. Polystyrene-pendant Ru-bipyridyl complex

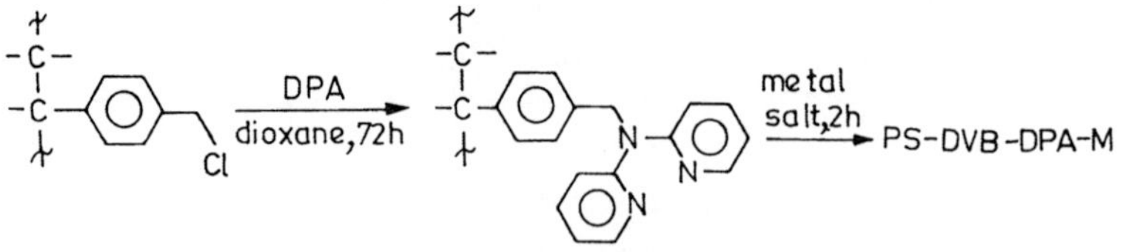

Fig. 9. Polystyrene-divinylbenzene-bipyridylamine metal complex

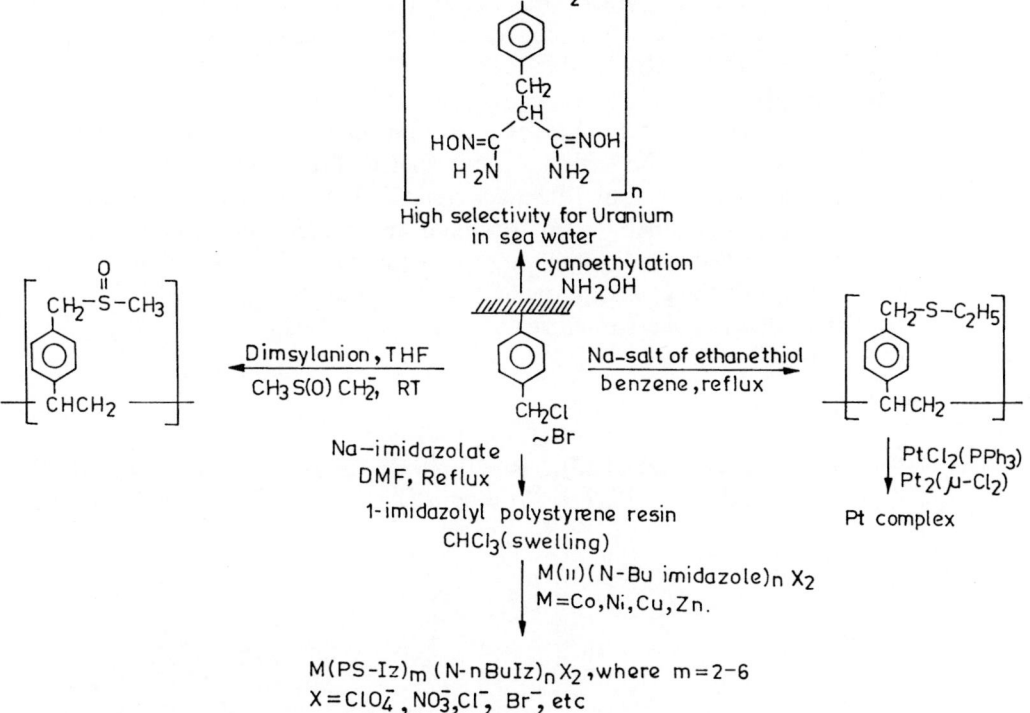

Fig. 10. 1-Carboxyl 8-naphthoyl substituted polystyrene

Fig. 11. Polystyrene-immobilized miscellaneous ligands

Miscellaneous ligands so far incorporated into the polystyrene matrix include imidazolyl [21], 2-amido oxime [22], thioether and sulfoxide moieties [23]. The procedure is simple and essentially involves the halogen-replacement of Cl- or Br- from polystyrene matrix, bearing $\sim CH_2Cl$, or $\sim Br$- groups. Mostly, first row transition element salts undergo facile complexation with these modified polystyrene matrices [21] (Fig. 11).

A novel photostabilization process of anchoring ligands onto a polystyrene support which would eliminate the possibility of leaching the metal moieties by the coordinating solvent is due to Cais et al. [24]. A typical system is the phenanthrene chromium dicarbonyl moiety, anchored onto polystyrene support by a photochemical ligand substitution reaction in which one of the CO moieties would be replaced by a phosphino ligand covalently bound to the polymer [24]. This would prevent the leaching of the Cr-moiety into the solution during a catalytic reaction in presence of the metallopolymer catalyst (Fig. 12).

Saegusa et al. [25] have developed a novel procedure of cationic grafting of phosphine oxide onto polystyrene. The 2-phenyl-1,2-oxaphospholane undergoes facile cationic ring opening polymerization in presence of benzyl chloride (Fig. 13). By analogy, a chloromethylated polystyrene (as the base polymer) readily yields a graft copolymer with a phosphine oxide graft chain with d.p. 4.7–10.5. This copolymer can be used as a chelating resin (bed type use) for UO_2^{2+}, Th^{4+}, Hg^{2+}, Pd^{2+} and Cu^{2+} to the extent of 80–100% depending on pH.

A novel cross-linking technique for localizing segments of polymeric chain molecules to give a three dimensional network structure displaying rubber like elasticity has been developed by Eichinger et al. [26]. An emulsion polymerized (isoprene-styrene/chloro-methylated styrene) copolymer on reaction with sodium salt of acetylacetone (acac) undergoes controlled ligand complexation (1.4×10^{-4} mol/polymer g). Subsequent cross-linking by chelation of acac groups with palladium acetate affords a highly elastomeric network [26].

2.2.3 Heterocyclic Polymer Systems

Ru(II)-poly(4-vinylpyridine) [27], Ru(II)-poly(6-vinyl-2,2′-bipyridine) [28], Ru(II)-poly(4-methyl-4′-vinyl-2,2′-bipyridine) [28, 29], Cu(II)-poly(4-vinyl-

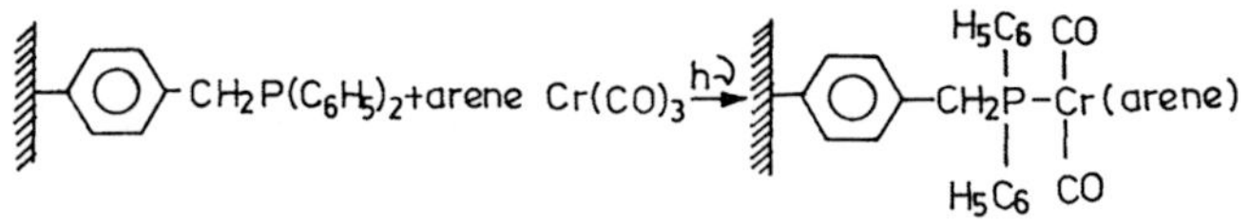

Fig. 12. Polystyrene anchored arene-chromium carbonyl complex

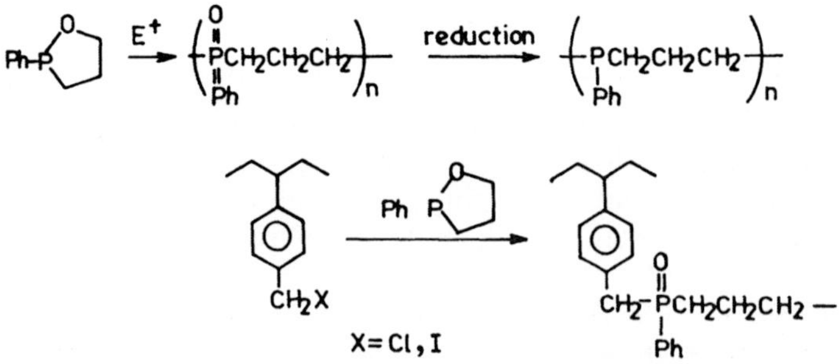

Fig. 13. Cationic grafting of phosphine oxide onto polystyrene

pyridine) cross-linked with 1,4-dibromobutane [30], Cr-complex supported polyvinylpyridine resins [31], Cu(II)-[(2-methyl-5-vinylpyridine)-co-acrylic acid] [32] represent some of the widely studied systems as application-oriented materials. A typical synthesis is illustrated schematically (Fig. 14).

Tsuchida et al. [33], synthesized Poly(1-vinyl and 1-vinyl-2-methyl-imidazole) bound heme (iron-porphyrin) complexes as models for mimicking natural oxygen carriers such as hemoglobin. The reaction between AIBN polymerized poly(1-vinyl)imidazole, or poly(1-vinyl-2-methylimidazole) and Fe(III) protoporphyrin IX diethyl ester in DMF and small amount of $Na_2S_2O_4$ yields the Fe(III) complex (Fig. 15).

N-Vinylpyrrolidone polymers are of special interest in medicine in the process of detoxification as well as for binding and removal of undesirable metallic ions and known as chelatotherapeutic agents [34]. Free radical co-polymers of poly(N-vinylpyrrolidone) and copolymer of N-vinylpyrrolidone and vinylacetate, vinylamine, vinylamidosuccinic acid are known to bind Cu^{2+} and other transition metal ions, and the resultant complexes exhibit interesting physico-chemical properties.

A polymeric ligand having pendant sulfide and imidazolyl groups binds Cu^{2+} to give an efficient oxidation catalyst [35]. The base polymer is a free radical copolymer of ethylvinylsulfide and vinylimidazole.

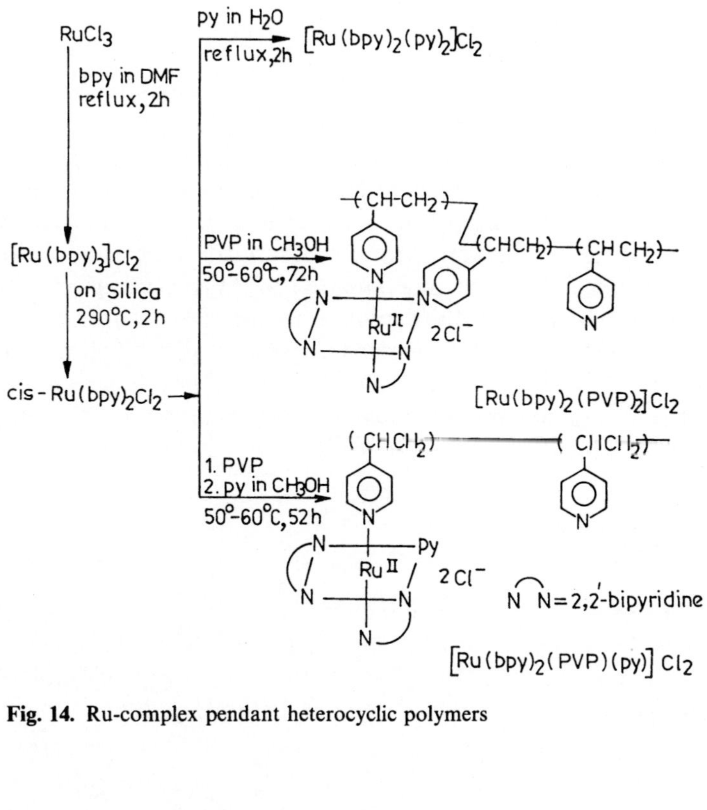

Fig. 14. Ru-complex pendant heterocyclic polymers

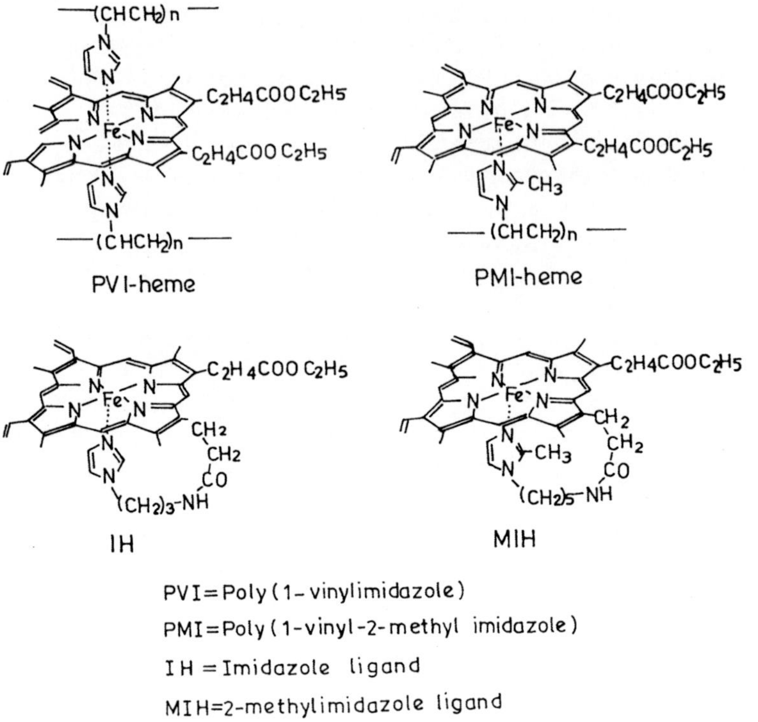

PVI-heme

PMI-heme

IH

MIH

PVI = Poly (1- vinylimidazole)
PMI = Poly (1-vinyl-2-methyl imidazole)
IH = Imidazole ligand
MIH = 2-methylimidazole ligand

Fig. 15. Polymer bound iron porphyrine complexes

Template Polymers. Template effects in chelating polymers constitute an interesting development in the field of metal containing polymers. The 'Template effects' are interpreted by the fact that the small molecule is templating a pattern in the macromolecule which can be recognized by the same molecule in a subsequent process. The idea is to prepare a polymer from the metal-chelated monomer, to remove the metal ion, and then to measure the selectivity of the prepared polymer for the metal ion of the template [36]. Typical examples of template systems are 4-vinyl-4'-methylbipyridine (Neckers [36]) and 1-vinyl-imidazole (Tsuchida [37]). These are polymerized in presence of divinylbenzene [36] and appropriate metal salts (Co^{2+}, Cu^{2+}, Ni^{2+}, Zn^{2+}). The template metal ions are removed by acid leaching and the polymer subsequently used for metal ion absorption studies (Fig. 16).

2.2.4 Halogeno-Telechelics

Halogeno-telechelic polymers result from complete ionization of both ends of telechelic prepolymers. Teyssie et al. [38] developed a facile process of synthesizing dicarboxylato polymers based on Group IVb metal ions (Ti/Zr). Addition of a stoichiometric amount of tetra-*n*-butoxy-titanium Ti(O-nBu)$_4$ to a carefully

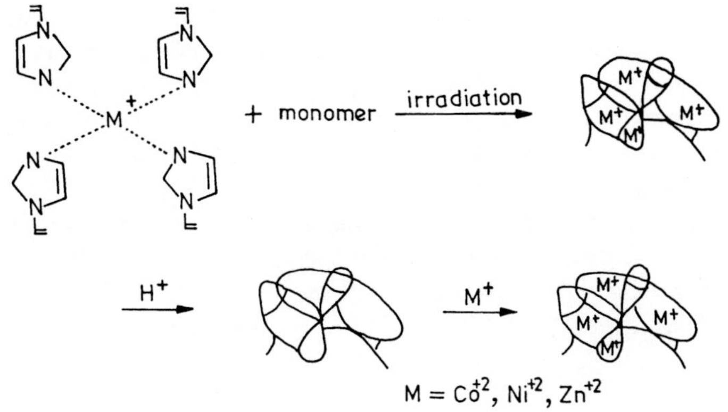

Fig. 16. Template polymers

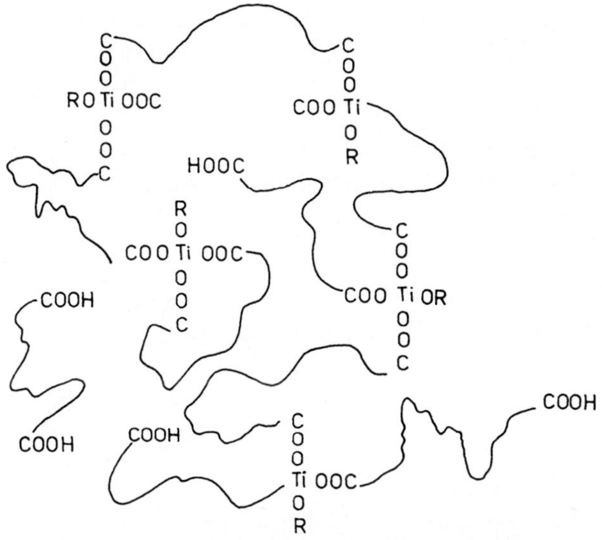

Fig. 17. Addition product of tetra-*n*-butoxy-titanium to carboxy telechelic polybutadiene

dried 5% solution of a carboxy telechelic polybutadiene results in Ti-complex schematically shown in Fig. 17. Halogeno-telechelic polyurethane-ureas have recently been synthesized from divalent metal salts (Mg^{2+}, Ca^{2+}) of *p*-amino-benzoic acid, diamine, dialkyl glycols and diisocyanates [39].

2.2.5 Miscellaneous Polycondensates

Polyamides. Metallized plastics [40] have been obtained from metal chelates of nylon 4 and nylon 6 by soaking a film of the polyamides cast from formic acid/propanol mixture into a metal salt solution followed by reduction with NaBH$_4$. In this context, Huang et al. [41] developed an unique technique of 'retroplating out' in which activated metal plate (powder) surfaces are brought in contact for 5–10 seconds at room temperature with polyamide or polyvinyl-alcohol or polyacrylamide metal chelate films to induce the retroplating out process.

Poly(amido-amine)s. Poly(amido-amine)s [42] offer a system to correlate the effect on Cu(II) complex forming ability with the number (n) of methylene groups between aminic nitrogen. The polymers possess the structures of the type shown in (Fig. 18) and compounds with n = 4 fail to form Cu(II)-complexes in aqueous solution.

Polyimides. In situ co-deposition of metal salts such as Co(II), or LiCl into a polyimide precursor from 3,3′,4,4′-tetracarboxybenzophenone dianhydride and 4,4′-diaminobenzophenone with subsequent thermal curing offers surface-conductive polyimide film [43]. By similar procedures, Taylor et al. prepared a series of polyimides modified with Pd, Pt, Ag, Au, Cu, Sn, Ti and magnetic-Fe [44, 45, 46].

Polyurethanes. Metal-containing polyurethanes have been synthesized by poly-condensing Co^{2+}, Cu^{2+} salts of mono(hydroxyethyl) phthalate with hexa-methylene diisocyanate and tolylene diisocyanates [47] (Fig. 19).

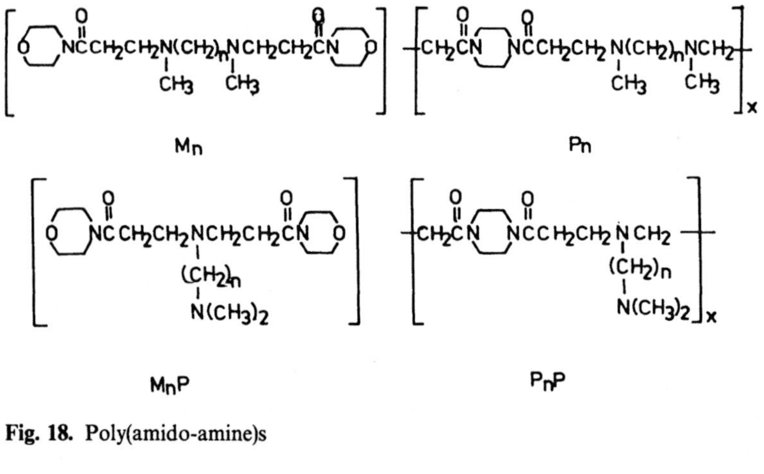

Fig. 18. Poly(amido-amine)s

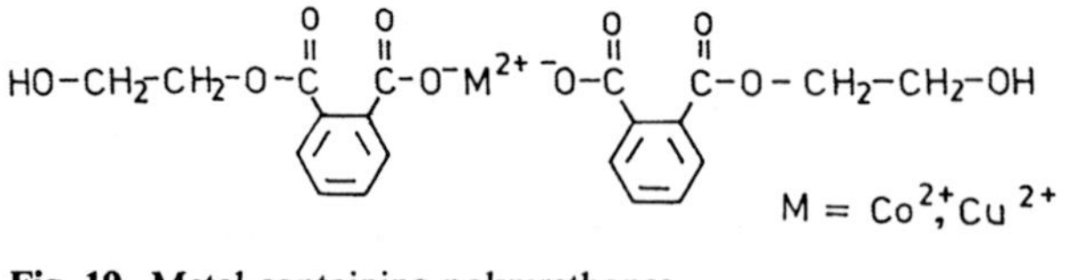

Fig. 19. Metal-containing polyurethanes

Polycondensates with the Ligand as a Comonomer. Metal-bearing polycondensates in which the ligand moiety alone or in combination with the coordinated
metal ions form one of the backbone comonomers are exemplified by the
following polymers reported by Wudl et al. [48], Biswas and Mazumdar [49,
50] and by Takahashi et al. [51] in the synthesis of organotin polymer by a
similar condensate (Figs. 20, 21).

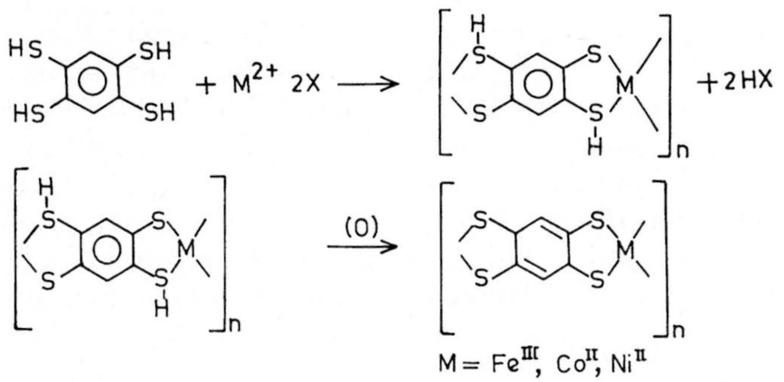

Fig. 20. Metal-containing polycondensates

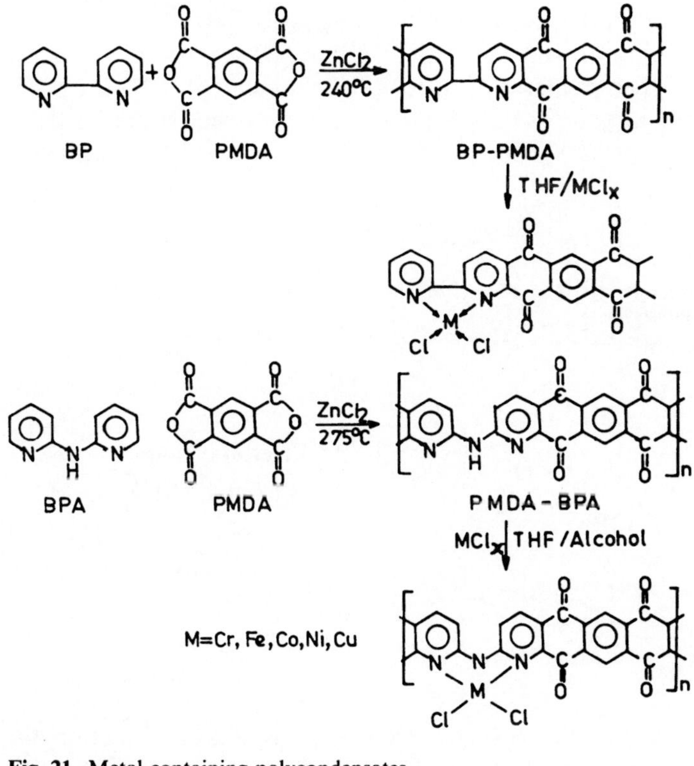

Fig. 21. Metal-containing polycondensates

2.2.6 Poly(ethylene)oxide

Poly(ethylene)oxide-salt complexes have of late received considerable attention in view of their use as solid polymeric electrolytes. Complexes of PEO with $LiBF_4$ [52], $LiCF_3SO_3$ [52], modified with polydimethyl siloxane [53], $LiClO_4$ [54, 55], Li^+, Na^+, K^+, Mg^{++} and Ba^{++}, NaI [56], NaSCN [56], poly-(2-sulfonethyl methacrylate-Li) [57], poly(2,4-dicarboxyhexafluorobutyl-ethoxy methacrylate-Li salt) [57] have been thoroughly investigated.

The PEO salt complexes are generally prepared by direct interaction in solution for soluble systems or by immersion method, soaking the network cross-linked PEO in the appropriate salt solution [52–57]. Besides PEO, poly(propylene)oxide, poly(ethylene)succinate, poly(epichlorohydrin), and poly(ethylene imine) have also been explored as base polymers for solid electrolytes [58]. Poly(ethylene imine) (PEI) is prepared by the ring-opening polymerization of 2-methyloxazoline. Solid solutions of PEI and NaI are obtained by dissolving both in acetonitrile (80 °C) followed by cooling to room temperature and solvent evaporation in vacuo. Polyethyleneimine-$NaCF_3SO_3$ complexes have also been explored [59].

Phosphate -ester cross-linked poly(ethylene glycol)s [60] are obtained from the condensation of glycols with $POCl_3$. Partition of Li-trifluoro-methanesulfonate-$LiCF_3SO_3$ between acetone solution and the polymer gel results in the formation of the electrolyte salt complexes [60].

Ti and Zr containing polytetramethylene oxide (PTMO) ceramic hybrid materials have lately been prepared by a sol-gel technique [61, 62]. Trialkoxy silane capped organic oligomer (PTMO or polyarylene ether sulfones) back-bones with titanium isopropoxide or Zr-(n-propoxide) are used in this process:

Sol formation:

$$M(OPr)_4 \xrightarrow[\text{isopropanol}]{HCl/H_2O} OPr-\underset{\underset{OPr}{|}}{\overset{\overset{OPr}{|}}{M}}-(O-\underset{\underset{OPr}{|}}{\overset{\overset{OPr}{|}}{M}}-)-O-\underset{\underset{OPr}{|}}{\overset{\overset{OPr}{|}}{M}}-Or$$

gelation:

$$OPr-\underset{\underset{OPr}{|}}{\overset{\overset{OPr}{|}}{M}}-(O-\underset{\underset{OPr}{|}}{\overset{\overset{OPr}{|}}{M}}-)-O-\underset{\underset{OPr}{|}}{\overset{\overset{OPr}{|}}{M}}-OPr \rightarrow \quad \text{Triethoxysilane capped PTMO}$$

$$\rightarrow \quad M - PTMO - Network.$$

2.2.7 Metal Porphorines

Polymer-bound metalloporphorines are receiving considerable attention because of their various desirable properties. They are available by polymerization of vinyl group-containing porphorine or by immobilization of suitable porphorines on reactive polymers. The latter method has the obvious advantage that complicated preparations are avoided. Some of the frequently used systems in this regard are macroreticular resins in which porphorin moieties are monomolecularly dispersed, Hasegawa et al. [63], Yamakita and Hayakawa [64] avoided the use of macroreticular resins by graft copolymerizing Cl-Me-styrene onto porous polyolefins and subsequently making ester linkages between Cl-Me groups and the carboxyl groups of metalloporphorins. Some typical systems are 4-Cl-methyl styrene + N-vinyl pyrrolidone and their metal derivatives through radical copolymerization of styrene with Fe(II) or Fe(III) protoporphorins [65], and Cl-methylated polystyrene bound with combined Zn-complexes of a phthalocyanine [66] and a porphorin moiety [66] (Fig. 22).

Water-soluble polymer-bound porphorins have been prepared by the reaction of poly(methacrylic acid) and poly(N-vinyl pyrrolidone-co-methacrylic acid) with the low-molecular weight substituted Zn complexes: tetraphenylporphorin, phthalocyanins and naphthocyanins [67, 68]. These reactions are

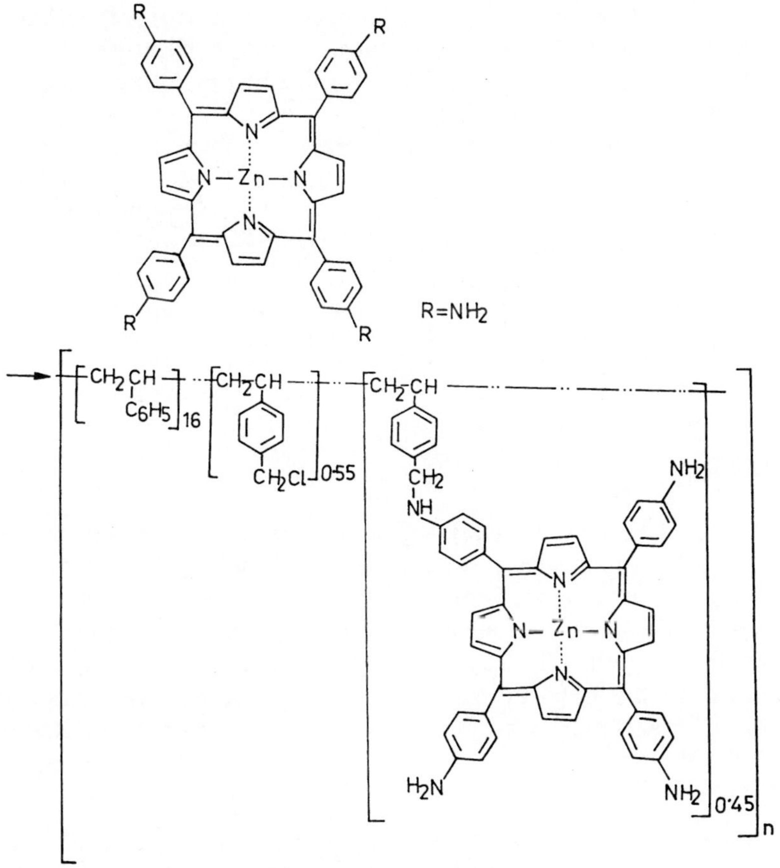

Fig. 22. Polystyrene-anchored Zn-tetraphenylporphorin

conveniently performed by dissolving the base polymers in DMF/acetonitrile and subsequently activating the CO_2H groups by addition of triphenylphosphine, tetrachloromethane and triethylamine. This solution readily reacts with the appropriate porphorin moiety to form the charged polymer (containing residual unreacted COOH groups) or 'uncharged' ones (with the residual COOH groups being converted into methylester with diazomethane). Positively charged water-soluble styrene based polymer bound porphorins have also been prepared through the reaction of poly(chloromethyl styrene) with low molecular weight porphorins in presence of triethylamine [68].

2.2.8 Metal Phthalocyanine-Imide Polymers

Poly(metal phthalocyanine)imide copolymers are produced from the reaction of metal (II) 4,4',4'',4'''-phthalocyanine tetramine (Cu, Co, Ni), diamines: 4-phenylene, 4,4'-bis(4-aminophenyl)methane, 9,9'-bis(4-aminophenylfluorene) and 1,2,4,5-benzene tetracarboxylic dianhydride [69].

A novel synthetic approach to preparation of conductive polyphthalocynins[1] is due to Lin and Dudek who employ thermal cyclization of poly (Cu(II), 2,3,9,10,16,17,23,24-octacyano)phthalocyanine to obtain a conducting system with extended π-conjugation [70].

Shirai et al. [71] also synthesized a novel class of polyimides containing M(II) phthalocyanine rings by solution condensation in N-Me-2-pyrrolidone of M(II) [2,9, or 10,2,16 or 17-bis(3,4-dicarboxy benzoyl)] phthalocyanine dianhydride with 2,6-diaminopyridine followed by thermal imidation [71] (Fig. 23).

Wohrle et al. [66] reported the synthesis of covalently bound phthalocyanine moieties to chloromethylated polystyrene polymethacrylic acid poly(N-vinylpyrrolidone-co-methacrylic acid) [67].

2.2.9 Crown Ethers

Poly(crown ether)s have received a great deal of attention in view of their cation-binding selectivity. They are expected to show a higher binding ability for large alkali metal cations than the monomeric analogues due to a cooperative action of two neighbouring crown ether moieties in the polymer chain. The cooperative action can be affected by the relative position of the neighbouring crown-ether moieties attached to the polymer backbone [72, 73]. In a novel synthetic procedure, Shirai et al. [73] have developed a facile way to fix the relative

---

[1] In line with the scope of this article only developments in polymer immobilized metal phthalocyanines, Schiff's bases, or crown ethers are highlighted barring a few cases of unanchored metal complexes displaying novel features.

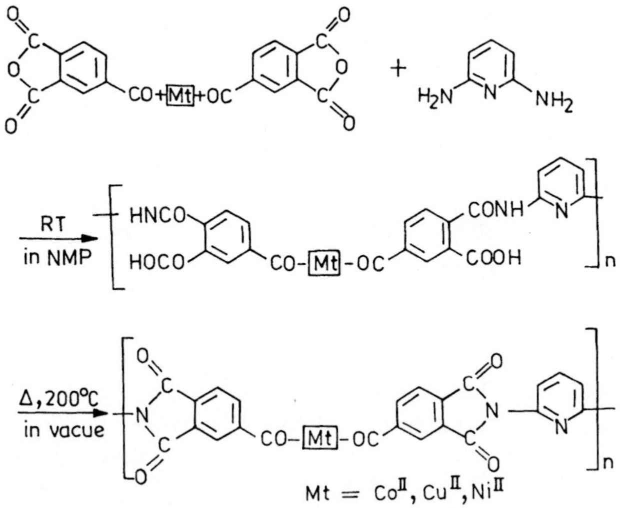

Fig. 23. Metal-phthalocyanine-containing polyimide

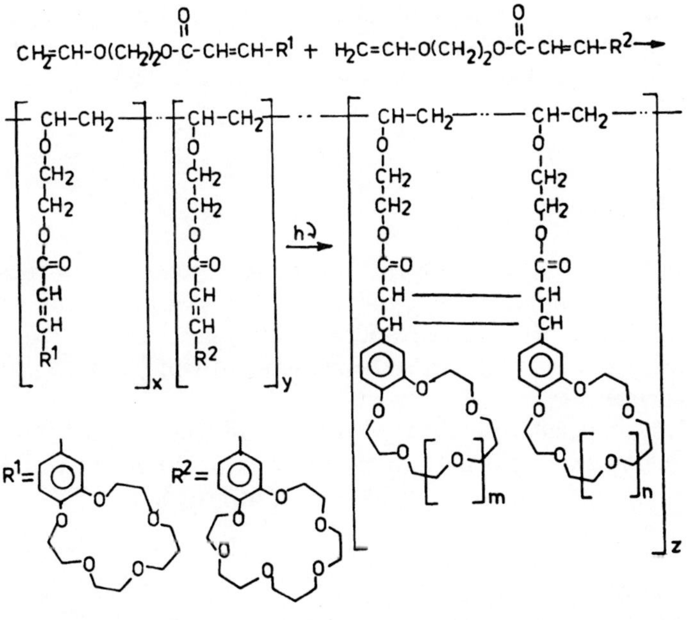

Fig. 24. Photodimerization of cinnamoyl group bearing polymer pendant crown-ethers

position of two crown ether moieties by photodimerization of cinnamic acid moieties introduced in the monomer. Thus, a benzo 15-Crown-5 having ether acryloyl or vinylic functionality undergo crosslinking through photodimerisation through the unsaturated moieties. (Fig. 24).

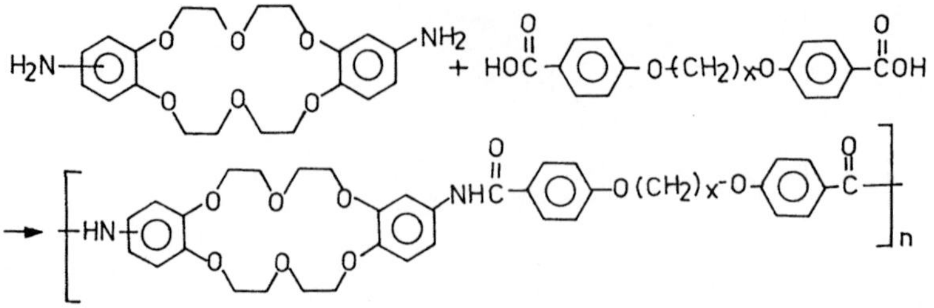

Fig. 25. Crown-ether-bearing polyamides

Crown-ether network polymers have also been prepared by adding to a slurry of NaH in THF, an equimolar amount of the appropriate hydroxy-containing crown-ethers followed by the addition of chloromethylated cross-linked polystyrene. Binding of Na, K, Cs picrates and of sodium tetraphenyl-borate to the immobilized crown-ether is achieved under simple conditions [74].

A new class of polyamides containing dibenzo 18-crown-6-moieties and alkaline units in the main chain shows [75] varying complexing capabilities with 4-toluene sulfonates of Rb^+, K^+, Na^+ and potassium salts of $CH_3C_6H_4SO_3^-$, SCN^-, I^-, Br^-. The polyamide polycondensates are obtained from cis- or trans-4,4'-diaminodibenzo 18-crown-6-(DAC) and a 4,4'-dicarboxy-α, ω-diphenoxy-alkane (DCA) (Fig. 25).

2.2.10 Schiff's Base Chelates

Polymer bound Schiff's bases are receiving increasing attention, as these materials are expected to combine the selective properties of low molecular weight Schiff's base chelates with the advantage of polymer immobilization. Wohrle et al. [7] synthesized a copolymer of 2-butylimino-4-vinylphenol with S to prepare N,N,O-chelates [7].

N-alkylation of low molecular weight N_3O_2-ligands with chloromethylated polystyrene results in immobilized Schiff's base structures [76] (Fig. 26). Co(II), Mn(II) and Fe(II) complexes of these bases have been investigated [76].

2.2.11 Metal-Containing Liquid-Crystal Polymers

A Cu^{2+} containing linear polymer [77] exhibiting smectogenic behaviour is obtained by a polycondensation technique (Fig. 27). A homologous set of polycondensates of 4,4'-[1,12 dodecanediyl bis(oxy)] bis benzoic acid with bis[N-[[2,4 dihydrophenyl]methylene]-alkylamino] Cu(II) exhibits monotro-pic liquid crystalline behaviour with 4–13 carbon atoms in the alkylamino group [78].

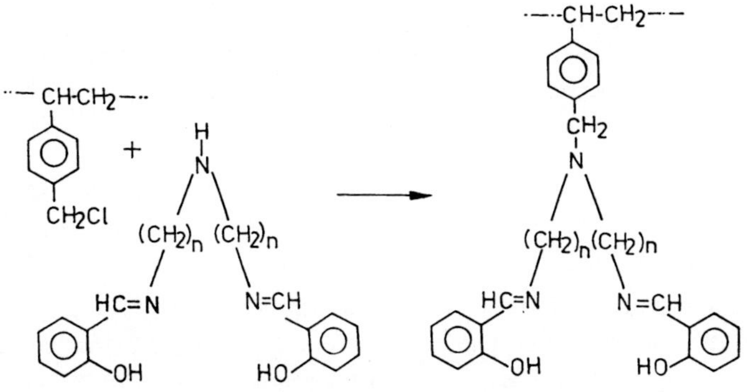

Fig. 26. Polymer-immobilized Schiff's base chelates

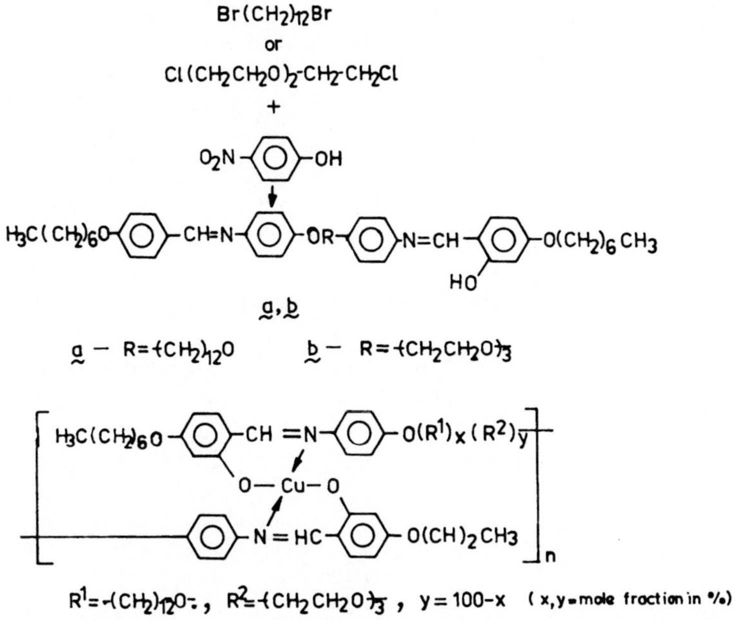

Fig. 27. Metal-containing liquid crystal polymer

2.3 Metal-Containing Plasma Polymers

The introduction of metals into plasma polymers has been the subject of some research during recent years. Basically, three methods have been used (a) Plasma polymerization of organometallic compounds [79, 80], (b) Concomitant etching or sputtering of metals [81] and (c) Evaporation of metals into a [82] plasma polymerization system. Munro et al. [83, 84] incorporated mercury into plasma-polymerized perfluorobenzene by mixing Hg-vapor with the perfluoro-benzene prior to entering them in the plasma reactor. Munro et al. also prepared [83, 84] plasma polymers of ferrocene, vinylferrocene and dimethylaminomethyl ferrocene.

2.4 Doping

Direct incorporation of metal ion moieties in a polymer chain by doping is a frequently used method. Mostly, acetylene polymers [85–88], poly(arylene vinylenes) [89], poly(phenylene sulfides) [90] with systems of conjugated π-electrons are doped directly with numerous metal salts and elements via charge transfer complex formation. Various metal salts used include, $PtCl_4$, $H_2PtCl_6 \cdot 6H_2O$, $LiClO_4$, $FeCl_3$ Ga, In, $TiCl_3$, and organo Li compounds.

Electrochemical doping of insulating polymers has been attempted for polyacetylene, polypyrrole, poly-N-vinyl carbazole and phthalocyaninato-poly-siloxane. Significantly, Shirota et al. [91] claim to have achieved the first synthesis of electrically conducting poly(vinyl ferrocene) by the method of electrochemical deposition (ECD) [91]. This is based on the insolubilization of doped polymers from a solution of neutral polymers. A typical procedure applied [91] for polyvinyl ferrocene is to dissolve the polymer in dichloromethane and oxidize it anodically with Ag/Ag^+ reference electrode under selective conditions. The modified polymer [91] (Fig. 28) is a partially oxidized mixed valence salt containing ferrocene and ferrocenium ion pendant groups with ClO_4^- as the counter anion.

A few systems obtained by the ECD method comprise polyacetylene tetra-chloroferrate [87] $[CH(FeCl_4)_y]_x$ and tetrachloroaluminate $[CH(AlCl_4)_x]$, Li^+ doping ($LiClO_4$ in propylene carbonate) [86] or $[(CH)_x(SbF_6)_y]_n$ (anodic oxidation in a solution of $[CH_3(CH_2)_3]_4 N^+ SbF_6^-$ in dry CH_2Cl_2) [92].

2.5 Mechanochemical Synthesis

Mechanochemical syntheses of some macromolecular complexes of Mn with polyamides and polyesters have been achieved. Simionescu et al. [93] reported the ultrasonic mechanochemical condensation of poly(ethylene terephthalate,

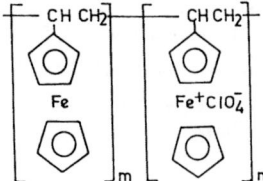

Fig. 28. Polyvinyl ferrocene

PET) with ethylenediamine as a ligand for V^{3+}, and of poly(ε-caprolactam) as a ligand for Mn^{2+}, by vibratory milling.

A compression moulding technique [94] has been used to prepare composites of polystyrene with layered perovskites $C_{12}Mn$; polystyrene powder (Mn = 145 000) and finely powdered $C_{12}Mn$ (average particle size 150 μm) are dryblended and subsequently compression moulded at 160 °C and a pressure of 50 kg/cm².

3 Evaluation of Metal-Containing Polymers

One of the obvious objectives of metal ion incorporation into polymers is to modify the essential bulk properties of these polymers with regard to miscellaneous end-uses. To this end however, structure-property correlations although useful, are difficult to achieve, particularly when high polymer networks are involved. This section will highlight how metal-ion incorporation can affect thermal stability, electrical and other useful properties of the polymer systems of which they are part.

3.1 Thermal Stability

Acrylic Polymers. Burrows et al. [95] showed by the 'Integral Procedural Decomposition Temperature' (IPDT) method that for main group metal ions – the stabilizing effect in regard to polyacrylamide is inversely proportional to the radius of the metal ion reemphasizing that the strength of the complex between the ion and the polymer is of importance in deciding the stability.

Polyvinyl Chloride. It is well-known how various attempts have been made to stabilise PVC against dehydrochlorination by salts, usually of divalent metal ions – as long chain alkylcarboxylates of Cd(II), Ba(II), Zn(II) [96]. Biswas and Moitra [102] recently established that the 3d metal ions incorporated in PVC-DMG-complex enhance the thermal stability of PVC in the order:

PVC-DMG-Co(II) > PVC-DMG-Ni(II) > PVC-DMG-Cu(II) > PVC

which is the same order as the 'crystal field stabilization energy' (CFSE) of the individual metal complexes.

Styrene Polymers. Polystyrene-divinylbenzene immobilized 3d metal bipyridylamine complexes also exhibit enhanced thermal stability as in the case with the polyacrylics. Biswas and Mukherjee [18] more recently reported the enhancement of thermal stability of 3d metal ion loaded PS-DVB-BPA-M(II) complexes in the order: $-Co(II) < -Fe(III) < -Cu(II)$. However, the $T_g S$ of these polymers as determined by a thermo mechanical analyzer, do not reveal any major variation from each other: PS-DVB (123 °C), PS-DVB-BPA (~), PS-DVB-BPA-Fe(II) (123 °C), PS-DVB-BPA-Co(II) (134 °C), PS-DVB-BPA-Cu(II) (120 °C).

Polyethylene Oxide and Related Polymers. Wright reported that NH_4^+, K^+ and Na^+ thiocyanates form crystalline complexes with PEO whose melting temperatures increase with decreasing cation size ($T_m = 343, 373$ and 468 K with NH_4^+, K^+, Na^+ respectively) [56]. The complexation is believed [56] to involve coordination of ethereal oxygen atoms to the cation as in macrocyclic polyether complexes. Complexation with NaSCN of a poly(ethylene-oxide-b-isoprene-b-ethyleneoxide) [97], occurs selectivity with PEO end blocks and yields a semicrystalline thermoplastic elastomer melting at 450 K, 90 K lower than that of the complexed block polymer. The heterogeneous nuclei are segregated in a few isolated crystallizable microdomains and hence cannot contribute to the nucleation of the greater part of the crystallizable component.

DSC traces of polyethyleneimine complexed with NaI also give evidence [58] for strong interaction of the salt with the polymer by a decrease in crystallinity of the polymer and at a molar (NaI/CH_2CH_2NH) ratio of 0.15, crystallization is completely prevented. At higher concentrations, a new endotherm manifests at 150 °C due to the specific crystalline complex between NaI and PEO. DTA studies of PEO-NaI and PEO-NaSCN complexes further reveal that melting temperature of the lamellar phase (crystalline) is independent of the nature of the anion [56].

Poly-condensates. Taylor et al. observed [98] that high temperature adhesive properties of the polyimide derived from 3,3',4,4'-tetracarboxybenzophenone dianhydride (BTDA) and 3,3'-diaminobenzophenone (DAB) are significantly enhanced by doping with $Al(AcAc)_3$. The polyimide (BTDA + DAB) similarly [98] yields on doping with $[(n\text{-}Bu)_3 PCuI]_4$ a flexible thin film. Copper increases the softening temperature of the polyimide in air with optimization being reached at around 2.6% Cu. Cu-weight percent upto 2.6% has little effect on the decomposition temperature of the polymers [98]. XPS studies of $[(n\text{-}Bu)_3 PCu]_4$ dopant during thermal imidization reveal that the dopant exhibits no copper satellite structure (indicative of the presence of Cu(I)). Upon curing the film containing this additive, Cu(I) is oxidized to Cu(II) on the film side exposed to oxygen, while the film surface on the glass side shows complete retention of Cu(I) spectrum. The effect of Cu(II) dopants as 1,3-diketonates and Schiff's base derivatives on imidization of BTDA + DAB – is to induce a higher apparent T_g and low-temperature weight losses (Cu(I) dopant > Cu(II)) even though, total decomposition occurs at a higher temperature in Cu(I) than in Cu(II) system. Taylor et al. [43] also confirmed that in situ co-deposition of Co and Li salts in condensation polyimides results in lower stability than the same obtained by Li/Co ion deposition [43] – both, however, causing lower stabilities in the parent polyimide. This may be due to (1) Co/Li may either act synergistically to reduce the stability or (2) higher overall ionic content in the codoped sample vs the singly doped film may be the factor.

Mechanochemically introduced V^{3+} and Mn^{2+} complexes of PET also exhibit enhanced thermal stability of the base polymer – an observation in line with the usual expectation [93].

Biswas and Mazumdar [49, 50] reported a similar enhancement of thermal stability on metal ion incorporation for the poly-condensate PMDA-BP/BPA/M, (Fig. 2) with the following features of interest: (1) With a typical metal ion incorporated in either PMDA-BP or PMDA-BPA, initial decomposition temperature of the metal-loaded polymers does not change significantly: PMDA-BP (238 °C) PMDA-BPA (235 °C), PMDA-BP-Fe(III) (280 °C), PMDA-BPA-Fe(III) (290 °C), PMDA-BP-Cu(II) (265 °C), PMDA-BP-Ni(II) (265 °C), PMDA-BP-Ni(II) (263 °C), PMDA-BPA-Ni(II) (280 °C). (2) With either PMDA-BP or -BPA, the effect of different metal ions on the stability is in the order: $Fe^{3+} > Ni^{2+} > Cu^{2+}$ (upto 4% decomposition) in the BPA complex, and upto 25% in the BP complex. (3) Beyond this temperature, the order in stability becomes same in either system: $Fe^{3+} > Cu^{2+} > Ni^{2+}$.

Phthalocyanine Polymers. Phthalocyanin-imide polymers show an initial decomposition temperature > 500 °C both in air and inert atmosphere (Co, Ni, Cu, Zn) as expected. An increase in the concentration of metal phthalocyanine in the copolymer increases the thermal stability [70]. Poly(Cu 2,3,9,10,16,17,23,24-octacyanophthalocyanine) represents an unique polymer showing enhanced thermal stability (1.2% wt loss at 585 °C and 1.5% wt loss at 625 °C, 21.6% at 800 °C) in He atmosphere: Rapid oxidation takes place on heating above 560 °C (9% wt loss at 585 °C) [99] in air. The enhanced stability of this material is different from that of monomeric metal phthalocyanine compounds which sublime and loose most of their weight around 600 °C [99].

Schiff's Bases. Coordination polymers of Ag(I), Cu(II), Zn(II) and Ti(IV) exhibit improved thermal stability in the order Cu < Ag < Ti < Zn [100]. Replacement of a benzyl by a phenyl group as substituent at the amino group of the ligand also enhances the stability.

3.2 Electrical Properties: Conductivity and Dielectric Characteristics

Acrylic Polymers. PMMA films doped with 1.7% wt $FeCl_3$ exhibit [101] a characteristic log conductivity vs $1/T$ plot comprising two segments separated by a glass-rubber transition temperature ($E_g = 25$ kcal/mol; $E_p = 16.3$ kcal/mol respectively). For the doped system however, three segments manifest. Activation energies for the 1st and 3rd segments decrease with increasing $FeCl_3$

concentration implying that FeCl$_3$ enters the matrix through attachment via the ester group and thereby assists the formation of continuous conduction paths in the polymer matrix. SEM and electron microprobe analyses endorse that the FeCl$_3$ doped PMMA contains three phases α, β and γ; the β phase being the halide-rich phase distributed in the least halide containing phase (α) and the highest halide containing γ phase. Characteristically, E$_2$ (β phase) increases upto 4% FeCl$_3$ implying increase of charge carrier density and falls subsequently at higher FeCl$_3$ concentration.

Polyvinyl Chloride. Biswas and Moitra [102] observed substantial increase in conductivity for metal modified PVC (Fig. 29). Table 1 presents the electrical conductivity data of the PVC-DMG-M(II) complexes. Interestingly, conductivities appreciably increase relative to PVC in the order PVC < PVC-DMG-Cu(II) < PVC-DMG-Ni(II) < PVC-DMG-Co(II). The enhancement in the conductivity is readily ascribable to the varying extents of charge transfer between the 3d metal ion centers and the electron-rich heteroatoms in DMG. Apparently, ease of such charge transfer will depend upon the availability of 3d vacant orbitals which follows the order Co^{2+} (3d^7) > Ni^{2+} (3d^8) > Cu^{2+} (3d^9).

Dielectric Characteristics. Fig. 30 reveals that at low frequency (10 kHz-100 Hz) the permittivities obey the trend PVC < PVC-DMG-Cu(II) < PVC-DMG-Ni(II) < PVC-DMG-Co(II). The permittivities fall monotonously with the applied frequency. The dielectric loss (tan δ) parameter also falls steadily with the applied frequency. However, PVC-DMG-Co(II) exhibits a comparatively large fall in tan δ from 2.5 to 0.2 in the frequency range (100 KHz-10 MHz), while

M = Fe^{III}, Co^{II}, Ni^{II}, Cu^{II} **Fig. 29.** PVC-DMG-M Polymer

Table 1. Electrical conductivity of PVC-DMG-M(II) [102]

Polymer	Diameter (mm)	Thickness (mm)	Voltage (V)	Current (A)	Electrical conductivity (Ω^{-1} cm^{-1})
PVC (pure)	11.6	1.75	20	0.14 × 10^{-8}	1.68 × 10^{-11}
PVC-DMG-Cu	11.8	1.85	20	6.1 × 10^{-8}	5.44 × 10^{-10}
PVC-DMG-Ni	11.65	1.55	20	0.6 × 10^{-6}	4.36 × 10^{-9}
PVC-DMG-Co	11.75	2.00	20	9.0 × 10^{-4}	8.30 × 10^{-6}

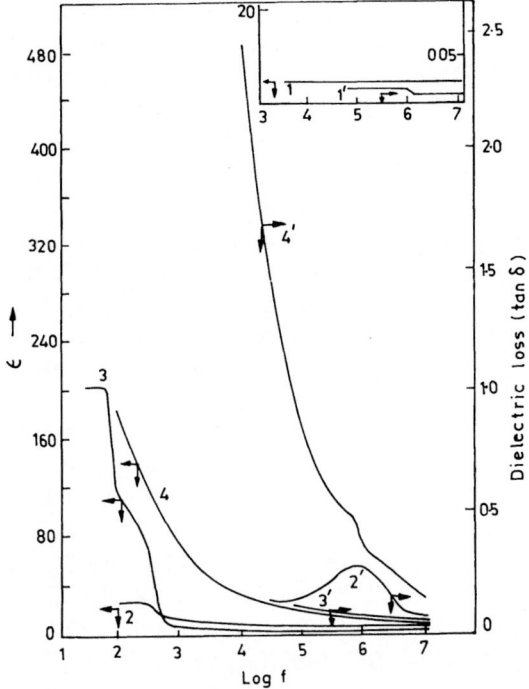

Fig. 30. Dielectric constant and tan δ versus frequency curve for PVC-DMG-M complexes (1,1′PVC; 2,2′PVC-DMG-Cu(II); 3,3′ ~ Ni(II), 4,4′ ~ Co(II).)

PVC-DMG-Ni(II) shows a smooth fall in tan δ from 0.1 to 0.05 in the same frequency range. Notably, tan δ frequency curves for PVC-DMG-Ni(II) or PVC-DMG-Co(II) do not reveal any maxima at the point of inflection in the dielectric constant frequency curve, Further, tan δ_{max} does occur for PVC-DMG-Cu(II) at a frequency which does not correspond to the inflection point in the permittivity-frequency curve. The introduction of polar groups in PVC will cause them to orient when placed in an electric field. If these groups are flexibly attached to the polymer chain, they will orient easily and rapidly. If the polymer is rigid, and the polar groups are rigidly attached, they will orient slowly with difficulty. In an alternating electric field, the polar groups in the polymer will orient and give high permittivity only when the frequency of alteration is low enough to permit motion and orientation of these groups. With increasing frequency in the alteration of electric field, the polar groups will be able to orient less and less rapidly and at still higher frequencies, they will be able to orient hardly at all. As a result the polymer will exhibit low permittivity. As for the broad nature of loss tangent parameter, it seems that loss of electric energy by conversion to thermal energy is too small to be detected at low frequency. It shows up only at some intermediate frequency in the transition region, where the polar groups are able to orient at the rate and frequency of alteration in the electric field.

Styrene Polymers. While polystyrene shows [103] a room temperature dc conductivity in the order $3 \times 10^{-20} \, \Omega^{-1} \, cm^{-1}$, the complexes of polystyrene with metallic salts, PS-AlCuCl$_4$, PS-AgAlCl$_4$, PS-AlCl$_3$, PS-CuCl and PS-AgClO$_4$ exhibit values in the order 10^{-11}–10^{-16}. The enhancement in the conductivities arises from larger extents of charge-transfer interaction between aromatic nucleus of polystyrene and the complexing metal salts, simple salts (AgClO$_4$) exhibiting lower values than the double-salts.

Volume resistivity [94] of polystyrene compression moulded with organo-metallic $[(n\text{-}C_{12}H_{25}NH_3)_2 MnCl_4]$ layered perovskites fall in the range 10^{10}–$10^{18} \, \Omega \, cm$ depending upon (1) the percentage of the filler, the effect being more pronounced at 80 °C than at 30 °C, and (2) orientation of the layers in the compression moulded structure. Such anisotrophy in conductivity is also known for nylon 6–6 and other systems.

Polyacetylenes. Numerous attempts have been made to enhance electrical conductivity of modified polydiacetylenes by doping. Both 'n' and 'p' type doping has been studied. Organo-lithium compounds have been widely used as dopants. Salts of transition metal ions WCl$_6$, MoCl$_5$, H$_2$IrCl$_6 \cdot 6H_2O$, FeCl$_3$, are receiving increased attention as dopants for polyacetylenes. Table 2 collects some significant data to compare the conductivities in polyacetylenes in the presence of metallic dopants. A mechanistic understanding of doping in poly-acetylenes has involved a variety of surface/structure characterization tech-niques like X-ray diffraction, scanning electron micrography [88], transmission electron microscopy (TEM) [88], Raman and X-ray photoelectron spectroscopy [88] and casting microprobe analysis [104]. XPS results show that upon doping, oxidation of polyacetylene chain is accompanied by partial reduction of Pt^{4+} to Pt^{2+}, the dopant anion being PtCl$_6^{2-}$ for both PtCl$_4$ and H$_2$PtCl$_6$, 6H$_2$O – doped polyacetylene [85]. SEM studies of metal halide doped poly-acetylenes have confirmed fibrillar nature of (CH)$_x$ films and decrease of porosity due to swelling of fibrils [88]. Interestingly, it has been endorsed by EPR, IR/visible and conductivity measurements that by doping with Li-benzophen-one [86], *cis*-rich (CH)$_x$ films can reach the metallic level, and Raman spectro-scopy confirms that doping is followed by isomerization of (CH)$_x$ from *cis* to *trans* structure [86].

Table 2. Conductivities in polyacetylenes in presence of metallic dopants

Polymer	Dopant	Conductivity ($\Omega^{-1} cm^{-}$)	Ref.
Polyacetylene	Li-benzophenone	> 100	86
-do-	PtCl$_4$	134	85
-do-	H$_2$PtCl$_6$, 6H$_2$O	12	85
-do-	RhCl$_3$	6×10^{-4}	85
-do-	CuCl$_2$	1.8×10^{-3}	85
-do-	MoCl$_5$	200	104
-do-	WCl$_6$	200	104

Phthalocyanine Polymers. As mentioned earlier [99], thermal cyclization of poly(Cu,2,3,9,10,16,17,23,24-octacyanophthalocyanine) induces a dramatic improvement of the conductivities. The polymer, cyclized at 203 °C, has a room temperature conductivity of *ca*, $6.7 \times 10^{-6} (\Omega^{-1} \text{cm}^{-1})$ in air; cyclized at 400 °C, the conductivity is enhanced 3 times which is claimed to be due to elimination of impurities. At 600–700 °C, large extensive cyclization (both inter and intramolecular) leads to highly conjugated structure; cyclized at 900 °C, the polymer exhibits conductivity in the range $4.6–8 (\Omega^{-1} \text{cm}^{-1})$ with only *ca.* 21.9% weight-loss.

Polycondensates. Miscellaneous copolycondensates with ligands as comonomers bind metal ions to produce conducting materials. Table 3 collects a few available data on such systems.

Dielectric Characteristics. The dielectric constant frequency data [49, 50] for PMDA-BP, PMDA-BPA and their metal-complexes are presented in Figs. 31 and 32 respectively and suggest the following typical features: (1) The dielectric values for PMDA-BP and PMDA-BPA complexes fall between 7–8, which is in

Table 3. Conductivities in some polycondensates

Polycondensate	Metal ions	Conductivity $(\Omega^{-1} \text{cm}^{-1})$	Ref.
Poly(metal tetrathiooxalates)	Cu^{II}, Ni^{II}, Pd^{II}	1–20 Stable on exposure to ambient environments for several months	105
Pyromellitic dianhydride-bipyridyl/ bipyridyl amine	Cu^{II}, Ni^{II}, Fe^{II} Fe^{III}, Cr^{III}	1.7×10^{-8} $- 5 \times 10^{-10}$ depending upon metal ion	49, 50
Polyarylenevinylenes	AsF_5 doped	$1.8–3 \times 10^{-7}$ depending on substituents	89

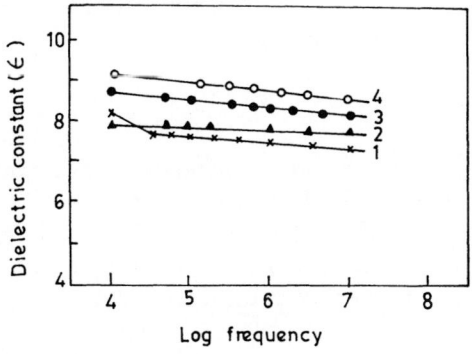

Fig. 31. Dielectric constant frequency data for PMDA-BP/PMDA-BP-M complexes: 1. PMDA-BP; 2, ~ Cu(II); 3, ~ Fe(III); 4, ~ Cr(III).

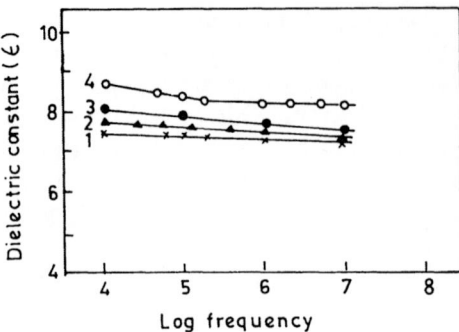

Fig. 32. Dielectric constant frequency data for PMDA-BPA/PMDA-BPA-M complexes: 1, PMDA-BPA; 2, ~ Cu(II), 3, ~ Fe(III), 4, ~ Cr(III)

the range expected for polar polymers. Changing the frequency from 10^3 to 10^7 Hz alters the dielectric values by only 3%. (2) On incorporation of metal ions in these complexes, the dielectric constant values are slightly enhanced in the above frequency range. (3) At a particular frequency (ca. 10 KHz) the dielectric constant values for the PMDA-BP-M complexes depend slightly on the nature of the metal ion in the order:

$$PMDA\text{-}BP\text{-}Cr(III)\ (9.2) > PMDA\text{-}BP\text{-}Fe(III)\ (8.7)$$

$$> PMDA\text{-}BP\text{-}Cu(II)\ (8.1) > PMDA\text{-}BP\ (8.0)$$

A similar trend is also noted for the PMDA-BPA-M complexes. (iv) These polymers exhibit rather low dielectric loss (tan δ: $1\text{–}4 \times 10^{-2}$ at 10 KHz which however is strongly dependent on the applied frequency. Thus, a broad relaxation pattern is typically observed even though the dielectric constant is not drastically changed. The dielectric polarization in the BPA/BP-metal complexes, is not evidently appreciable since the conjugation between the two pyridyl rings in BP or BPA is not favoured structurally. The system is best described as a rubi-conjugated system which apparently should discourage manifestation of high dielectric constant and high conductivity. The observed order in dielectric constant with the 3d-metal ions is significantly also the order in which the electronegativity of the metal ions changes:

$$Cr^{3+}\ (1.56) < Fe^{3+}\ (1.64) < Co^{2+}\ (1.70) < Cu^{2+}\ (1.78)$$

This implies that the electronegativity difference between nitrogen and the metal decreases in the series leading to a decreased extent of dielectric polarization as actually observed. The frequency dependence of the tan δ values in these complexes is marked. Evidently, the metal complexes possess a rigid structure where dipoles do not find sufficient time to reorient with the direction of applied frequency of alteration resulting thereby in a broad dielectric relaxation.

Table 4. Conductivities[a] of polyethylenic and allied systems

System	Mol. wt	Complexed with	Conductivities ohm^{-1}cm^{-1}	Temp. K range	Ref.
PEO	600	NaI (1:6)	10^{-3}–10^{-6}	298–450	56
	10 000	NaI (1:4)	10^{-3}–10^{-8}	298–450	56
	600 000	NaI (1:3.2)	10^{-5}–10^{-9}	298–450	56
	10 000	NaSCN	10^{-3}–10^{-9}	298–450	56
PEO	400	LiBF$_4$ (s)[b]	10^{-3}–10^{-5}	298–450	52
	400	LiBF$_4$(us)[c]	3.1×10^{-4}–3.1×10^{-5}	298–450	52
	5×10^6	LiBF$_4$	10^{-5}–10^{-7}	298–450	52
	5×10^6	LiCF$_3$SO$_3$	10^{-5}–10^{-7}	298–450	52
PEO	4×10^6	CH$_3$COOLi[d]			
		O/L$^+$ = 4	3.16×10^{-4}–10^{-9}	298–420	57
		O/Li$^+$ = 9	10^{-4}–10^{-9}	298–420	57
		O/Li$^+$ = 18	10^{-4}–10^{-8}	194–420	57
		CH$_3$SO$_3$Li			
		O/Li$^+$ = 9	10^{-3}–10^{-7}	298–420	57
		O/Li$^+$ = 18	10^{-3}–10^{-7}	298–420	57
		CF$_3$CoLi	10^{-3}–10^{-9}	298–420	57
		CF$_3$CF$_2$COLi	10^{-3}–10^{-8}	298–420	57
		CF$_3$CF$_2$CF$_2$COOLi	10^{-4}–10^{-9}	298–420	57
		LiHFG[e]	10^{-4}–10^{-10}	298–420	57
PEO	5×10^6	Na$^+$ with TCNQ[f]	10^{-2}–10^{-4}[g]	298–420	
PEO (cross-linked)	3000	LiClO$_4$ LiClO$_4$/EO unit = 0.2	10^{-2}–10^{-9}	253–373	55
PEO (linear)	3000	LiClO$_4$/EO unit = 0.05	10^{-8}–10^{-11}	253–373	55
Phosphate ester cross-linked	400	LiCF$_3$SO$_3$			
Polyethylene glycols		O/Li$^+$ = 27.6	5.2×10^{-6}	293	60
		O/Li$^+$ = 13.7	2.5×10^{-6}	293	60
Poly (ethylenimine) (PEI)	2×10^5	[(EI)$_x$NaCF$_3$SO$_3$]$_n$			
		x = 4	5.6×10^{-8}	314	59
			1.2×10^{-5}	367	59
		x = 5	2.4×10^{-7}	314	59
			5.4×10^{-5}	366	59
		x = 6	2.4×10^{-7}	314	59
			5.4×10^{-5}	366	59
PEI	2000	NaI 0.3 mol. ratio	10^{-3}–10^{-8}	340–450	58
PEO*	400	LiClO$_4$	10^{-4}–10^{-7}	20–50%	54

[a] ac conductivities are reported if not mentioned otherwise.
[b] Saturated concentrations (s).
[c] Unsaturated concentrations (us).
[d] Salt concentrations are quoted as the ratio of the number of moles of oxygen atoms in PEO to the number of moles of Li$^+$ ions (O/Li$^+$).
[e] Dilithium hexafluoroglutarate LiOOC(CF$_2$)COOLi (LiHFG).
[f] Tetracyanoquinodimethane radical anions (TCNQ).
[g] dc Conductivity reported.
* Conductivity measured as a percentage (%) of epoxyglycidylether of bisphenol A

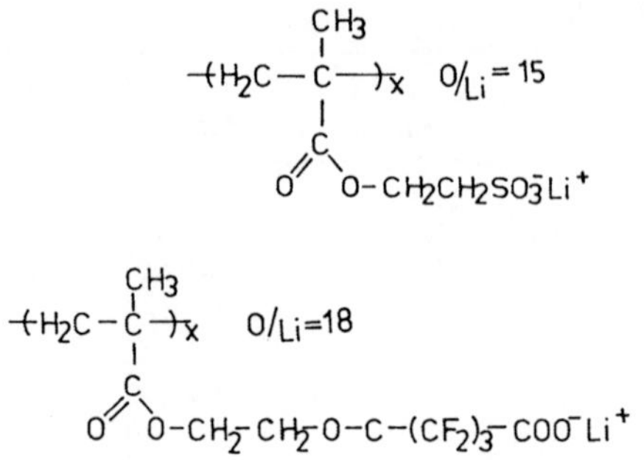

$$+H_2C-\underset{\underset{\underset{O^{\diagup\diagdown}O-CH_2CH_2SO_3^-Li^+}{|}}{C}}{\overset{\overset{CH_3}{|}}{C}}\rightarrow_x \quad O_{Li}=15$$

$$+H_2C-\underset{\underset{\underset{O^{\diagup\diagdown}O-CH_2^-CH_2^-O-C-(CF_2)_3^-COO^-Li^+}{|}}{C}}{\overset{\overset{CH_3}{|}}{C}}\rightarrow_x \quad O_{Li}=18$$

Fig. 33. Blend of PEO with the Li-salt of anionic polymer

PEO and Related Systems. High ionic conductivities have been characteristically associated with polymer-alkali metal complexes, which are receiving great deal of research attention as electrolytes for solid state batteries. $LiClO_4$ dispersed homogeneously in cross-linked (β-cyanoethyl methylsiloxane); poly(β-cyanoethyl methylsiloxane-*co*-dimethylsiloxane) shows a network film conducting in the order of 10^{-5} ohm^{-1} cm^{-1} at room temperature [106].

In alkali metal salt–PEO complexes [52, 54. 55], the alkali metal ions are situated in an electron-rich environment surrounded by ethereal oxygen of PEO and the charge-carriers are the cations and the anions. In Li-thiocyanate-PEO complex the transport number of Li^+ is 0.5, as determined by several workers. Ward et al. [57] have developed two simple procedures to achieve Li^+ transport numbers close to unity: (1) by incorporating anion as part of a polymer chain when their transport number will be zero and the cations will be the major charge carriers. This is achieved [57] by producing blends of PEO and Li salts of the ionic polymer as in Fig. 33 and by inhibiting the mobility of the anion by using a Li salt of a dibasic acid, dilithium hexafluoroglutarate.

More recently, Tsuchida et al. [107] achieved a Na^+ ionic conductivity of $1 \times 10^{-6}\,\Omega^{-1}\,cm^{-1}$ in a composite film made from Nafion (perfluorosulfonate ions) and diendo-acetylated polyoxy ethylene, due apparently to enhanced dissociation of the sodium perfluoro sulfonated groups in the composite matrix.

Table 4 summarizes the conductivity ranges reported for various PEO–salt systems and suggests that the conductivities depend on (a) polymer molecular weight (b) polymer–salt ratio (c) temperature. As expected, the polymer–salt ratio is one of the most significant factors.

4 Applications and Concluding Remarks

A few remarks would be in order on the potentiality of metal-containing polymers as application-oriented materials. The major applications of these materials are in the following directions: (a) electronic materials, anisotropic optical materials – active species in electronic energy-transfer processes of lasers (polymer matrices containing europium chelates) – 'Solar energy conversion' (polymer–pendant Ru bipyridyl complexes) [108], (b) solid state batteries (polyethylene oxide-alkali metal salt complexes) (c) heterogenized-homogeneous catalysts – oxidation catalyst displaying catalyst like activity [109, 110], selective hydrogenation catalysts for organic nitrocompounds [12], (Rhodium complex of polystyrene with anthranilic acid) [12], silica-supported poly-acrylonitrile), polymerization catalyst (PVC-DMG-M(II), Cu, Co, Ni-isotactic rich poly N-vinylcarbazole) [111].

New areas employing inorganic and organometallic polymers are being constantly explored. The interaction of metal moieties with natural product polymers as possible drugs is one potential field warranting in-depth research. Development of structurally-novel metal-containing polymers can face the challenge of providing new materials for applications as superconducting materials, ultra-high strength materials, liquid-crystals and bio-compatible polymers.

Acknowledgements. Thank are due to CSIR, New Delhi for a research grant to MB and the authorities of IIT Kharagpur for facilities.

References

1. Arimato FS, Haven AC (1955) J Am Chem Soc 77: 6295
2. Pittman CU, Sheats JE, Carraher CE, Zeldin M, Currel B (1989) Procs ACS Division of Polymeric Materials (Miami Beach Meeting) 61: 91; Biswas M, Moitra S (1986) Polym Prepr 27: 76.
3. Sharma PR, Tripathi KC, Gupta HM, (1980) J Polymer Sci, Polymer Chemistry Ed 18: 2609; Dharia JR, Pathak CP, Babu GN, Gupta SK (1988) J Polymer Sci, Polymer Chemistry Ed 26: 595
4. Itoh H, Kondo S, Masuda E, Hanabusa K, Sherai H, Najo N (1986) Makromol Chem, Rapid Communications 7: 585
5. Milovskaya EB, Zamoiskaya LV (1982) Polymer 23: 891
6. Gressiar JC, Levesque G, Patin H, Varret F (1983) Macromolecules 16: 1577
7. Wohlre D, Bohlen H, Meyer G (1984) Polymer Bulletin 11: 143
8. Kurimura Y, Vehino Y, Ohta F, Saito C, Koide M, Tsuchida E (1981) Polymer J 13: 247
9. Yamagishi A, Nitta K (1982) Polymer 23: 1177
10. Walsh DJ, Crosby P, Dalton RF (1983) Polymer 24: 423
11. Bai RK, Zong HJ (1984) Makromol Chem, Rapid Communications 5: 501
12. Guo XX, Zong HJ (1984) Makromol Chem, Rapid Communications 5: 507
13. Hirasawa E, Hamazaki H, Tadano K, Yano S (1991) J of Applied Polymer Sci 42: 621
14. Neckers DC (1985) In: Sheats JE, Carraher CE Jr, Pitman CU Jr (eds) Metal-containing polymer sytems. Plenum, New York, p 385
15. Phillips HH, Kenstle JF, Adoock JL (1981) J Polymer Sci, Polymer Chemistry Ed 19: 175
16. Kaneko M, Ochiai M, Kunosita K, Yamada A (1982) J Polymer Sci, Polymer Chemistry Ed 20: 1011
17. Kratz MR, Hendricker DJ (1986) Polymer 27: 1641
18. Biswas M, Mukherjee A (1992) J Appl Polymer Sci 46: 1453
19. Kaneko M, Vamada A, Tsuchida E, Kurimura YJ (1982) Polymer Sci, Polymer Letters 20: 593
20. Veba Y, Zhu KJ, Banks E, Okamoto Y (1982) J Polymer Sci, Polymer Chemistry Ed 20: 1272
21. Welleman JA, Hulsberger FB, Reedijk J (1981) J Makromol Chem 82: 785
22. Kise H, Sata H (1985) Makromol Chem 186: 2449

23. Davies AJ, Sood A (1983) Makromol Chem Rapid Communications 4: 777
24. Tatarsky D, Khon DH, Caie M (1980) J Polymer Sci 18: 1387
25. Kabayashi S, Suzuki M, Saegusa T (1983) Macromolecules 16: 1010
26. Ning YP, Mark EJ, Iwamoto N, Eichinger E (1985) Macromolecules 18: 55
27. Shimidzu T, Izaki K, Akai Y, Jyoda T (1981) Polymer J 13: 889
28. Sunil K, Furue M, Nozakura SH (1984) J Polymer Sci, Polymer Chemistry Ed 22: 3779
29. Furue M, Sunil K, Nazakura SH (1982) J Polymer Sci, Polymer Letters Ed 20: 291
30. Nishide H, Tsuchida E (1981) J Polymer Sci, Polymer Chemistry Ed 19: 835
31. Brunlet T, Gelbard G, Guyot A (1981) Polymer Bulletin 5: 145
32. Bekturov EA, Kudaibergenov SE, Sagitov VB (1986) Polymer 27: 1269
33. Tsuchida E, Nishide H, Yokeeyama H, Inove H, Shirai T (1984) Polymer J 16: 325
34. Bekturov EA, Kudaibergenov SE, Kanapyanova GS, Kurmanlaeva AA (1984) Polymer
    Communications 25: 220
35. Sato M, Shindo H, Kondo K, Takenioto K (1980) J Polymer Sci, Polymer Chemistry Ed
    18: 101
36. Gupta SN, Neckers DC (1982) J Polymer Sci Polym Chem Ed 20: 1609
37. Kato M, Nishide H, Tsuchida E, Sasaki T (1981) J Polymer Sci 19: 1803
38. Broze G, Jerome R, Teyssie P (1982) J Polymer Sci, Polymer Chemistry Ed 21: 593
39. Matsuda H, Takechi S (1991) J of Polymer Sci, Polymer Chemistry Ed 29: 83
40. Yen C-C, Huang CJ, Chang T-C (1991) J of Applied Polymer Sci 42: 439
41. Huang CJ, Yen C-C, Chang T-C (1991) J of Applied Polymer Sci 42: 2237
42. Barbucci R, Casolaro M, Ferrute P, Barone V (1982) Polymer 23: 148
43. Rancourt JD, Taylor LT (1987) Macromolecules 20: 790
44. Furstch TA, Taylor LT, Fritz TW, Fortner G, Khor E (1982) J Polymer Sci, Polymer
    Chemistry 20: 1287
45. Bergmeister, Rancaert JD, Taylor LT (1990) Chemistry of Materials 2: 640
46. Taylor LT, Rancourt JD (1990) In: Sheats J, et al. (eds) Inorganic and metal-containing
    polymeric materials. Plenum, New York, p 109
47. Kothandaram H, Venkatarao K, Raghavan A, Chandrasekaran V (1985) Polymer Bulletin
    13: 353
48. Dirk CW, Bousseau M, Barret PH, Heeger AJ, Wudl F (1986) Macromolecules 19: 266
49. Majumdar A, Biswas M (1991) Polymer Bulletin 26: 145
50. Majumdar A, Biswas M (1991) Polymer Bulletin 26: 151
51. Tahara T, Seto K, Takahashi S (1987) Polymer J 19: 301
52. Payne DR, Wright PK (1982) Polymer 23: 690
53. Albinsson I, Mellander BE, Stevens JR (1991) Polymer 32: 2712
54. Chiang CK, Baner BJ, Brider KM, Davis GT (1987) Polymer Communications 28: 34
55. Watanale M, Nagano S, Sanui K, Ogata N (1986) Polymer J 18: 809
56. Lee CC, Wright PV (1982) Polymer 23: 681 and references cited therein
57. Bannister DJ, Davies GR, Ward IM, Melntyre JE (1984) Polymer 25: 1291
58. Chiang CK, Davies GT, Harding CA, Takahashi T (1985) Macromolecules 18: 825, and
    references cited therein
59. Harris CS, Shriver DF, Ratner MA (1986) Macromolecules 19: 987
60. Giles RMJ, Greenhall PM (1986) Polymer Communications 27: 360
61. Wang B, Wilkes GL (1991) J Polymer Sci, Polymer Chemistry Ed 29: 905
62. Wang B, Wilkes GL, Hedrick JC, Leptak SC, McGrath J-E (1991) Macromolecules 24: 3449
63. Hasegawa E, Kanayama T, Tsuchida E (1977) J Polymer Sci, Polymer Chemistry Ed 15: 3039
64. Yamakita H, Hayakawa K (1980) J Polymer Sci, Polymer Letters Ed 18: 529
65. Finkenaur AL, Dickinson LC, Chien JCW (1983) Macromolecules 16: 728
66. Wohrle D, Krawczyk G (1986) Makromol Chem 187: 2535
67. Wohrle D, Krawczyk G, Paliuras M (1988) Makromol Chem 189: 1001
68. Wohrle D, Krawczyk G, Paliuras M (1988) Makromol Chem 189: 1013
69. Achar BN, Fohlen GM, Parker JA (1985) J Polymer Sci, Polymer Chem Ed 23: 801
70. Lin JWD, Dudek LP (1985) J Polymer Sci, Polymer Chemistry Ed 23: 1579
71. Shirai H, Takemae Y, Kobayashi K, Kondo Y, Hirabaru O, Hojo N (1984) Makromol Chem
    185: 1395
72. Shirai M, Orikata T, Tanaka M (1983) Makromol Chem Rapid Communications 4: 65
73. Shirai M, Veda A, Tanaka M (1985) Makromol Chem 186: 2619
74. Sinta R, Lamb B, Smid J (1983) Macromolecules 16: 1383
75. Jin J-II, Lee MS, Kim SJ (1984) Polymer J 16: 547

76. Aeissen H, Wohrle D (1981) Makromol Chem 182: 2961
77. Carfagana C, Caruso U, Roviello A, Sirigu A (1987) Makromol Chem, Rapid Communications 8: 345
78. Caruso V, Roviello A, Sirigu A (1991) Macromolecules 24: 2606
79. Inagaki N, Yagi T, Katsuura K (1982) Eur Polym J 18: 621
80. Inagaki N, Ohkuto J (1991) J of Applied Polymer Sci 43: 793
81. Kay E, Dilks A, Hetzler U (1979) J Macromol Sci Chem 12: 1393
82. Asano Y, (1983) Thin Solid Films 8: 1
83. Munro HS, Till C (1984) J Polymer Sci, Polymer Chem Ed 22: 3933
84. Munro HS, Eaves JG (1985) J Polymer Sci, Polymer Chemistry Ed 23: 507
85. Cao Y, Guo K, Wan M, Wang P, Qian R, Wang F, Zhao X (1983) Polymer Communications 24: 300
86. Rachdi F, Bernier P, Faulques E, Lefrant S, Schuel F (1982) Polymer 23: 173
87. Chen Z, Shen Z, Chen C, Zhao Y, Yang M (1987) Makromol Chem 188: 349
88. Begin D, Demai JJ, Vangeliste R, Billand D (1986) Polymer Communications 27: 117
89. Antoun S, Gaghon DR, Karasz FE, Lenz RW (1986) Polymer Bulletin 15: 181
90. Chance R, Shacklette LW, Eckhardt H, Sowa JM, Elsenbaumer RL, Ivory DM, MIller GG, Baughman RH (1981) In: Seymour RB (ed) Conductive polymers. Plenum, New York, p 125
91. Shirota Y, Kakuta T, Mikaya H (1984) Makromol Chem, Rapid Communications 5: 337
92. Makenbusch M, Weiners G (1983) Makromol Chem, Rapid Communications 4: 555
93. Simionescu C, Opera CV, Negulianu C (1980) Makromol Chem 181: 1579
94. Addeo A, Carfagha C, Nicodema L, Nicolais L (1983) J Polymer Sci, Polymer Letters Ed 21: 317
95. Burrows HD, Ellis HA, Utah SI (1981) Polymer 22: 1740
96. Nassa LI (1970) In: Mark HF, Gaylord NG, Bikales NM (eds) Encyclopedia of polymer science and technology, vol 12. Interscience, New York, p 725
97. Bobitaille C, Prud'honme J (1983) Macromolecules 16: 665
98. Ezzell SA, Furtsch TA, Khor E, Taylor LT (1983) J Polymer Science, Polymer Chemistry Ed 21: 865
99. Lin WP, Dudek LP (1985) J Polymer Sci, Polym Chem Ed 23: 1589
100. Grodzinski JJ (1980) Makromol Chem 181: 2441
101. Tawansi A, Soliman MA, Kinawy N, Badr SIM (1988) Polymer Bulletin 19: 289
102. Biswas M, Moitra S (1989) J Appl Polymer Sci 38: 1243
103. Hiirai H, Shigekura M, Komiyama M (1986) Makromol Chem, Rapid Communications 7: 351
104. Rolland M, Aldissi M, Schue F (1982) Polymer 23: 834
105. Reynolds JR, Lillya CP, Chien CW (1987) Macromolecules 20: 1184
106. Fang SB, Zhang P, Jiang YY (1988) Polymer Bulletin 19: 81
107. Takeoka S, Horiuchi K, Yamagata S, Tsuchida E (1991) Macromolecules 24: 2003
108. Kaneko M, Yamada A, (1985) In: Sheats JE, Carraher CE Jr, Pittman CU Jr (eds) Metal containing polymer systems. Plenum, New York, p 249
109. Shirai H, Maruyama A, Takano J, Kobayshi K, Hojo N, Urushinodo K (1980) Makromol Chem 181: 565
110. Shirai H, Maruyama A, Kobayashi K, Nobumasa H (1980) Makromol Chem 181: 575
111. Moitra S, Biswas M, Uryu T (1989) Polymer Communications 30: 225

Editor: Prof. Ledwith
Received October 1992

Author Index Volumes 101-115

Author Index Vols. 1-100 see Vol. 100

Subject Index

Springer-Verlag
and the Environment

We at Springer-Verlag firmly believe that an international science publisher has a special obligation to the environment, and our corporate policies consistently reflect this conviction.

We also expect our business partners – paper mills, printers, packaging manufacturers, etc. – to commit themselves to using environmentally friendly materials and production processes.

The paper in this book is made from low- or no-chlorine pulp and is acid free, in conformance with international standards for paper permanency.